Developing, Disseminating, and Assessing Command Narrative

Anchoring Command Efforts on a Coherent Story

WILLIAM MARCELLINO, CHRISTOPHER PAUL,
ELIZABETH L. PETRUN SAYERS, MICHAEL SCHWILLE,
RYAN BAUER, JASON R. VICK, WALTER F. LANDGRAF III

Approved for public release; distribution unlimited

NATIONAL SECURITY RESEARCH DIVISION

For more information on this publication, visit www.rand.org/t/RRA353-1

Library of Congress Cataloging-in-Publication Data is available for this publication.
ISBN: 978-1-9774-0684-2

Published by the RAND Corporation, Santa Monica, Calif.
© Copyright 2021 RAND Corporation
RAND® is a registered trademark.

Cover images: U.S. Marine Corps/Lance Cpl. Drake Nickels
Adobe Stock/Sergey Nivens; Cover design by Pete Soriano.

Support RAND
Make a tax-deductible charitable contribution at
www.rand.org/giving/contribute

www.rand.org

Preface

This publication outlines a set of research findings and recommendations for the production, dissemination, and assessment of command narrative. In addition to offering relevant concepts, definitions, and best practices from scientific research, the publication offers lessons learned from the joint force. While this research is primarily aimed at the geographic combatant commands, it will be of interest to any military command that needs to ensure that words and actions from the command are in harmony and support operational goals.

This research was sponsored by U.S. European Command and conducted within the International Security and Defense Policy Center of the RAND National Security Research Division (NSRD), which operates the RAND National Defense Research Institute, a federally funded research and development center sponsored by the Office of the Secretary of Defense, the Joint Staff, the Unified Combatant Commands, the Navy, the Marine Corps, the defense agencies, and the defense intelligence enterprise.

For more information on the RAND International Security and Defense Policy Center, see www.rand.org/nsrd/isdp or contact the director (contact information is provided on the webpage).

Contents

Figures and Tables

Figures

Tables

Summary

In the contemporary era marked by informational competition, one of the most important activities of a geographic combatant command (GCC) is the development, presentation, and support of the command's narrative. Recognizing the need for a theater-wide command narrative,[1] U.S. European Command asked RAND's National Defense Research Institute (NDRI) to conduct a study on developing, integrating, and assessing command narrative. While our study is directed at narrative at the GCC level, our recommendations have broader applicability for any command that operates at the strategic or operational level. To make the recommendations of the study as accessible as possible to practitioners, the project produced not only this report but also a four-page quick reference Smart Guide that provides definitions, summary guidance for developing a narrative, and a checklist for assessing the quality of the output of the narrative development process.[2]

What Is Command Narrative?

By *command narrative*, we mean the framework that anchors all the command's communication activities—everything from specific messages to exercises (because actions speak louder than words). Command narrative is broader than the commander's intent for a single operation. Command narrative comprises all operations, activities, and investments undertaken by the command and all public communication undertaken by the command. Command narrative is critical because it unifies and is supported by the various activities undertaken by and within the command. An effective command narrative can demonstrate resolve, bolstering the confidence of allies and partners and contributing to deterrence. Command narrative positions the desired end state as the ending of a connected set of stories, explains the role of military forces as characters, and gives a narrative explanation of "why" the end state matters and how it will be brought about by the characters.[4]

---

[1] Interview with experts in J-39, October 21, 2019.

[2] The Command Narrative Smart Guide (Marcellino et al., 2021) is available at www.rand.org/t/TLA353-1.

Research Approach

To better understand how the GCCs can most effectively use command narrative, we conducted two parallel research efforts: (1) a review of scholarly literature on narrative broadly as a means of persuasion and (2) site visits and interviews at the GCCs with military subject-matter experts (SMEs).

Scholarly Research on Narrative

Stories are powerful; humans "make sense, think, understand, and remember in specific story terms and elements" (Haven, 2014, p. 3). Before incoming information enters our awareness, it has already been processed and shaped into story form. This story takes the shape of what is known, or what makes sense to the individual, based on their lived experiences (Haven, 2014, p. 29). This filtering process is important to keep in mind, because humans will filter out what they do not know, understand, or otherwise counters their experiences.

Leveraging human cognition, narrative can engage individuals in much the same way that real-life experiences do. This is why narratives are so powerful, because people automatically perceive the world with mental models that affect both what they perceive and how they interpret that incoming information (Davidson, 2017, p. 2). According to the Joint Staff J-7 focus paper on communication strategy and synchronization, a compelling narrative helps to promote the legitimacy of the mission and "prevent the 'say-do' gap in which our actions and words conflict in the eyes of the audience" (Joint Staff J-7, Deployable Training Division, p. 1). Overall, our literature review provided several key findings relevant to the aspiring narrative practitioner.

Subject-Matter Expert Interviews and Geographic Combatant Command Site Visits

We conducted semistructured interviews with SMEs from the GCCs, as well as from the Marine Corps' Marine Expeditionary Force (MEF) Information Group (MIG) community.[3] We also spoke with individuals from the Office of the Secretary of Defense (OSD), the Joint Staff, functional combatant commands, and the military services. We also interviewed a range of military communication experts from the Public Affairs, Information Operations, and Psychological Operations specialties.

---

[3]   We note that the MIGs operate at the tactical and operational level within theater commands, and thus are not directly comparable to GCC commands. We think insights from MIGs practice and organization are highly relevant to this study, however. The U.S. Marine Corps is at the vanguard of integrating information as a joint function and service warfighting function, and the rest of the joint force can benefit from the Marine Corps' lessons learned.

Key Findings

Our SME interviews and site visits to the GCCs revealed many challenges to implementing command narrative. The majority of GCCs do not have command narratives, staff generally do not know what command narrative is, and existing processes and structures generally do not support the development, dissemination, and assessment of effective command narrative. Beyond command narrative, we found a wide variety of practices supporting communication efforts broadly, many of which seem to hinder effective command narrative.

However, we also found good models for command narrative use and many important best practices for communication efforts. Below, we have synthesized our findings from the scholarly literature and from practitioner interviews into a consolidated set of key findings.

Narrative matters because it is the primary cognitive framework human beings use to make sense of the world:

- Stories play an essential role in how people process information, and effective stories contain common core components.
- Narratives are powerful because they influence audiences.
- Defeating hostile narratives must go hand in hand with the promotion of alternative, positive narratives.
- Because it is hard to counter an accepted and engaging narrative, it is important to get the command's version of events out first.
- Actions are messages.
- Joint doctrine requires the GCC to develop a narrative for military operations, one that is synchronized with the U.S. national strategic narrative.
- Command narrative will be a priority to the degree that it is a commander's priority.
- Communication practices and emphases can be personality-dependent.

Effective narrative development must account for complexity—multiple concepts, moderators, and emotional strategies that make narrative more or less effective:

- It is important to understand the differences among key concepts, including messages, themes, stories, and narrative.
- Additional moderators—such as causality, temporality, coherency, character identification and proof, and plausibility—can be used to inform a story.
- Both positive and negative emotions can be effective and influential.

Audience is central to effective narrative:

- Audience understanding should remain central to narrative development and implementation.
- Other communication best practices, such as targeting, tailoring, and framing, can be leveraged in conjunction with narrative theory.

Effective communication practices, including command narrative, require assessment:

- Assessment demands time and resources but is necessary for determining whether communication efforts are making a difference.
- Good assessment requires good data practices.

There is wide variability in how combatant commands organize for and understand the broader issue of coherency in communications:

- Some commands fully integrate all communications through close collaboration and synchronization and anchors on the command's narrative, whereas others seek to deconflict essentially distinct communication efforts, without attention to command narrative.
- Intelligence has an important part to play in informing communication efforts.
- There are military cultural obstacles to effective communication efforts.
- There is no clear staff section or office at the GCC level responsible for narrative development.

Finally, communication efforts involve risk, and thus risk-aversion may hinder communication.

Recommendations

GCCs should consider how to leverage insights gleaned from our analysis. Taken individually or used in conjunction with one another, these recommendations can improve the development, implementation, and evaluation of narrative activities (including work related to messages, themes, and stories). While these are primarily framed around the GCCs, they could be adapted down to the component level, or up to multinational force levels (i.e., NATO). Drawing on our analysis of both lessons learned from interviews and site visits and building on the broader insights from our scientific literature around narrative and communication, we offer the following recommendations.

Systematically implement an evidence-based command narrative that supports the command's campaign objectives:

- Use narrative as a strategy to inform, engage, model behavior, persuade, and provide comfort (Shaffer et al., 2018, p. 434).
- Be prepared for dynamic change in using a narrative strategy.
- Leverage everyone in the command's potential to contribute to (or undermine) intended narratives or stories.
- Make command narrative and communication a priority.

Generate and maintain trust with audiences:

- Avoid the "say-do" gap.
- If using fictional elements in a narrative, make sure to do so clearly and intentionally.

Put the audience—not the command—at the center of narrative crafting:

- Identify and research existing narratives in the area of responsibility.
- Conduct formative audience research.
- Use audience segmentation to distinguish groups with different characteristics.

Implement best practices in crafting narratives:

- Nest narrative development from top to bottom.
- Use all eight story components when writing a story (to support a narrative).
- Create stories that transport individuals, such that the audience is "caught up in the story."
- Use additional moderators—such as causality, temporality, coherency, character identification and proof, and plausibility—based on desired story goals and objectives.
- Use additional best practices, such as positive or negative emotions, targeting, tailoring, and framing.
- Use assessment to make sure that communication is making a difference.

Implement best communication practices in military processes and structures:

- GCCs should
  - tear down the Public Affairs Office firewall
  - adopt a communications integration mindset and a close collaboration model
  - put out annual training guidance on narrative
  - build robust communication assessment capabilities

– use effects-based planning in narrative development
– prioritize commander involvement; effective use of command narrative requires attention and emphasis from the commander
– invest in narrative development. Effective narrative that produces effects will require a right-sized force
– designate a "keeper of the narrative" tasked with primary responsibility for narrative development.

- Narrative processes and structures must become institutionalized.

Our recommendations involve substantive, and potentially disruptive, changes but are necessary if GCCs wish to compete agilely and effectively. As we show in the study, everything commands do (and do not do) sends powerful messages that multiple audiences interpret using narrative frameworks. If GCCs and other commands with operational and strategic concerns wish to have campaign-relevant effects, they will need to be able to communicate in a coherent, coordinated, effective way that leverages humanity's storytelling nature, anchoring words and deeds in a harmonious way that enables the commander's intent.

Acknowledgments

We would like to thank COL Mike Jackson for his foresight in requesting this project and COL Ryan Keating and COL Robert Kjelden for their support and encouragement in researching effective command narrative practices while overseeing this project. We also thank Al Bal and Julie Weckerlein for their support and assistance as sponsor points of contact. We are further indebted to the host of personnel at several geographic combatant commands, other commands, and other defense organizations, both within the U.S. Department of Defense and among partner nations, who shared insights and experiences with us as subject-matter experts. We are prevented by the conditions of our interviews from thanking you individually and by name, but know that this effort would not have been possible without your participation and support.

Lastly, we would like to thank Phillip McGuinn and Todd Richmond for their invaluable advice and insight on this document as part of the quality assurance process.

Abbreviations

ALF	Animal Liberation Front
AWG	Asymmetric Warfare Group
CCS	commander's communication synchronization
CCSWG	commander's communication synchronization working group
COMSTRAT	communication strategy and operations
CSD	communication synchronization division
DoD	U.S. Department of Defense
EUCOM	U.S. European Command
GCC	geographic combatant command
IOWG	Information Operations Working Group
ISIS	Islamic State of Iraq and Syria
JWMC	Joint Military Information Support Operations (MISO) WebOPS Center
KLE	key leader engagement
KLEWG	Key Leader Engagement Working Group
MEF	Marine Expeditionary Force
MIG	MEF Information Group
NATO	North Atlantic Treaty Organization
NDRI	RAND National Defense Research Institute
OSD	Office of the Secretary of Defense
OSINT	open source intelligence
PA	Public Affairs
PAO	Public Affairs Office/Public Affairs Officer
SME	subject-matter expert
SOUTHCOM	U.S. Southern Command
StratComm	strategic communications

CHAPTER ONE

Introduction

In the contemporary era marked by informational competition, a critical activity of a geographic combatant command (GCC) is the development, presentation, and support of the command's narrative as part of the commander's communications synchronization (CCS) process. The CCS aligns communication regarding the joint force's mission with the broader national strategic narrative (Joint Publication [JP] 3-61, 2016, p. ix). Within this process, command narrative provides direction and guidance for communication on all levels, aligning "words and deeds."[1] Done well, a command narrative is supported by coherent themes and messages and unifies the various activities undertaken by and within the command: exercises, operations, and engagements. An effectively planned, executed, and supported command narrative can demonstrate resolve, bolstering the confidence of allies and partners and contributing to deterrence. In contrast, a poorly conceived, executed, and supported command narrative can harm GCC missions, leaving the command (or the United States) looking self-contradictory, inconsistent, or incomprehensible. Joint doctrine describes the requirement to develop command narrative, and ways to promulgate it, within the broader staff planning processes and particularly within the CCS. GCC staffs could benefit from reviewing joint doctrine and adapting it to their command's requirements.

Recognizing the need for a theater-wide command narrative,[2] U.S. European Command (EUCOM) asked the RAND National Defense Research Institute (NDRI) to conduct a practice-focused study on developing, integrating, and assessing command narrative. By *practice-focused*, we mean that while the study was informed by scientific research on narrative and communications broadly, the study was tightly focused on improving practice in the force: identifying best practices and principles that would inform day-to-day operations and planning at EUCOM. And while we started with EUCOM, our study looked broadly across the GCCs and also within and outside the joint force. We looked at how services such as the Marine Corps are tackling these challenges, how service institutions such as the war colleges understand them, and how

---

[1] Semistructured interview with personnel from the NATO Strategic Communications Centre of Excellence, in person, March 9, 2020.

[2] Interview with experts in J-39, October 21, 2019.

NATO approaches command narrative. Thus, while our study is directed at narrative at the GCC level, our recommendations have broader applicability for any command that operates at the strategic or operational level.

Additionally, we found that it was impossible to disentangle command narrative from persuasive communications more broadly. In many ways, effective command narrative entails good communications practices; for example, cycles of planning and assessment, or effective communications integration structures that dovetail with command narrative's harmonizing function. Thus, while this report is focused on command narrative, findings and recommendations often include broader communication best practices.

U.S. European Command's Need for Practice-Focused Improvement Regarding Narrative

EUCOM recognizes the importance of command narrative and the need to develop a robust capacity for communication that includes command narrative. As one expert in Joint Planning put it, the command understands the need to "operationalize messaging and synchronization so we can compete better" in the European theater.[3] Out of this need grew a major restructuring effort, moving from a commander's communication synchronization working group (CSSWG) within the J-39 to a communication synchronization division (CSD) within the J-5/8, reflecting the J-5 longer-term planning perspective.

Our study, then, is meant to directly inform and enable EUCOM's effort to establish a CSD that effectively marries different command operations to send a clear, unified message. Because the intent is effective competition, the command asked us to mine not just scientific and military literature on narrative and effective communications but, in particular, to look at actual practice, primarily at the GCC level. We were then able to base our analysis on a broad research endeavor around communication and persuasion, and to draw military-specific high-context insights to support our practice recommendations. To support EUCOM requirements for practice-focused advice, our efforts produced not only this report but also a four-page quick reference Smart Guide that provides definitions, summary guidance for developing a narrative, and a checklist for assessing the quality of the output of the narrative development process (Marcellino et al., 2021).[4]

---

[3] Group semistructured interview with experts in Joint Planning, in person, February 27, 2020.

[4] The Command Narrative Smart Guide is available at www.rand.org/t/TLA353-1.

What We Mean by *Command Narrative*

While we go into greater explanatory depth later in this report, by *command narrative* we mean the framework and overarching strategy that anchor all of the command's communication activities—everything from specific messages to exercises. Command narrative is a way of using a shared story for both internal and external audiences, to make sure there is no "say-do gap" (in which actions and words conflict in the eyes of the audience). For example, a command might have a narrative that starts with "The United States is a good neighbor, and a good neighbor . . ."—which is a powerful way to quickly let everyone know who the command is in this story and to leverage existing narratives about "what good neighbors do."

Command narrative is distinct from commander's intent. Again, while we more fully describe and define command narrative in Chapter Three, JP 3-0, *Joint Operations*, states that commander's intent is

> A clear and concise expression of the purpose of the operation and the desired military end state that supports mission command, provides focus to the staff, and helps subordinate and supporting commanders act to achieve the commander's desired results without further orders, even when the operation does not unfold as planned. (JP 3-0, 2017, p. II-7)

While commander's intent could be expressed in narrative form or include narrative elements, it is not synonymous with command narrative. Command narrative is broader than the commander's intent for a single operation; command narrative covers all operations, activities, and investments undertaken by the command and synchronizes all public communication undertaken by the command. Command narrative positions the desired end state as the ending of a story, explains the role of military forces as characters in the story, and gives a narrative explanation of "why" the end state matters and how it will be brought about by the story's characters (Paul, Colley, and Steckman, 2019).

Command Narrative and Doctrine

Although the concept of developing a narrative for military operations that is synchronized with the U.S. national strategic narrative within general military planning processes is well established in joint doctrine, our research revealed that it is not universally known among practitioners. We also found that practitioners at one GCC used terms such as *narrative, theme,* and *message* synonymously,[5] even though they

---

[5] Group semistructured interview with communication synchronization team (CST), in person, December 23, 2010.

are clearly and distinctly described in joint doctrine, as shown in Figure 1.1 (JP 3-61, 2016, pp. I-11–I-13).

As a result, we saw a perception that joint doctrine was lacking and that GCCs needed a systematic way of developing and promulgating narrative that goes beyond Public Affairs (PA) messaging.[6] Joint doctrine frames narrative within the CCS process to support planning and execution of a coherent national effort (JP 5-0, 2017, p. II-11) and frames the CCS within the information instrument of national power (JP 1, 2017, p. I-12). A non-authoritative joint doctrine note published in 2013 shares principles, techniques, processes, and best practices to aid joint force commanders and staffs in implementing communication synchronization (Joint Doctrine Note 2-13, 2013, p. i).

The process of developing a command narrative begins with identifying the U.S. national strategic narrative. Ideally, the President or National Security Council staff would create an explicit national strategic narrative at the outset of planning for specific military operations and include it with operational orders or other strategic guidance to the GCC. In the absence of an explicit narrative, GCC planners can derive the U.S. national strategic narrative from U.S. policy; the speeches of senior U.S. officials, such as the President and Secretary of State; and other national strategic documents (JP 3-61, 2016, pp. I-12–I-14). Identifying the U.S. national strategic narrative is an inherent GCC responsibility as it coordinates and synchronizes communications to ensure a coherent national effort in planning and executing military operations (JP 5-0, 2017, p. II-11).

GCCs use the CCS process to synchronize communications within the joint force and with coalition members, partners, host nations, adversaries, and other key

Figure 1.1
Narrative, Themes, and Messages

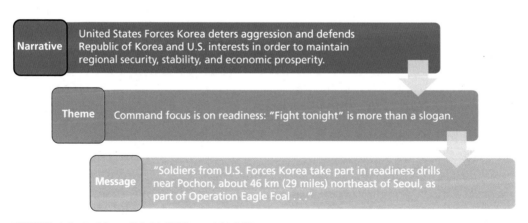

| Narrative | United States Forces Korea deters aggression and defends Republic of Korea and U.S. interests in order to maintain regional security, stability, and economic prosperity. |

| Theme | Command focus is on readiness: "Fight tonight" is more than a slogan. |

| Message | "Soldiers from U.S. Forces Korea take part in readiness drills near Pochon, about 46 km (29 miles) northeast of Seoul, as part of Operation Eagle Foal . . ." |

SOURCE: Adapted from JP 3-61, 2016, pp. I-11–I-13.

---

6 Semistructured interview with Plans and Strategy expert, in person, October 23, 2019.

audiences in their area of responsibility.[7] After developing a command narrative, GCC staffs must also develop supporting themes and messages. The *narrative* is a short story that underpins operations and provides greater understanding and context to U.S. and military operations (JP 3-61, 2016, p. I-11). A *theme* is a unifying idea that supports the narrative and is designed for broad application to achieve specific objectives (JP 3-0, 2011, p. III-16). A *message* is a focused communication directed at a specific audience to support a specific theme (JP 3-61, 2016, p. GL-5). The GCC staff develops narrative, themes, and messages during the overall military staff planning process to ensure alignment of GCC actions with U.S. national interests. The GCC makes adjustments as necessary by using the CCS as part of its continuous staff planning processes.

Beyond U.S. joint doctrine, NATO shares an essentially similar understanding. NATO's Strategic Communications Centre of Excellence defines *command narrative* as "a concise but comprehensive account of a situation, entailing the current state, a desired future state, a pathway to get there and a justification, based on a set of norms and values expressed through mediums such as stories or messages."[8] This definition includes four key elements in a NATO narrative: the current state, a future state, a pathway, and a justification.[9]

Research Approach

Our research effort followed two parallel lines of effort. We conducted a systematic review of scholarly literature around narrative broadly as a means of persuasion and as a basic way humans make sense of and interpret the world. Our aim was to review the broader scientific literature around persuasion and communication, grounding our effort empirically. We winnowed a very large set of potential research studies down to 90 that we deemed within scope and crucial for insight. From that review, we synthesized findings and drew out recommendations that support command narrative.

Our second line of effort was site visits and interviews at the GCCs, and with a variety of military subject-matter experts (SMEs) on command narrative. Ultimately, we conducted in-depth, semistructured interviews with 46 communication practitioners and SMEs across 31 organizations. These visits and interviews allowed us to understand how different organizations understand command narrative, and how different processes and structures within these organizations enabled (or failed to enable) command narrative development, dissemination, and assessment.

---

[7]   Depending on context, the term *key audiences* may include the general public and governments, as included in the U.S. Department of Defense (DoD) definition for "Audiences, Stakeholders and Publics" (JP 3-61, 2016, p. I-14).

[8]   Interview with personnel from the NATO Strategic Communications Centre of Excellence, March 9, 2020.

[9]   Interview with personnel from the NATO Strategic Communications Centre of Excellence, March 9, 2020.

Outline of This Report

In the rest of this report, we lay out our findings from the literature review, the site visits, and the SME interviews. In Chapter Two, we describe our method and summarize the data used for the literature review, and we explain various aspects of narrative, to inform the set of recommendations for the GCCs on effective use of narrative that we present in Chapter Four. In Chapter Three, we explore narrative use from a practitioner standpoint, explain how narrative is understood in the force, analyze various processes and structures for communication, and offer a set of key findings and attendant recommendations for the GCCs to develop, disseminate, and assess narrative. Chapter Four concludes the report, and Appendix A includes the interview protocol we used. Appendix B details the methods used to identify and select items for inclusion in our review of the literature. The key insights and guidance from this report are distilled and presented in an easy-to-use companion Smart Guide aimed at practitioners in the force, available at www.rand.org/t/TLA353-1.

CHAPTER TWO

Narrative Literature Review

Introduction

As part of our research into command narrative, we reviewed relevant scholarly literature, with a fairly broad aperture. In addition to command narrative, we also looked at research on narrative more broadly as a means of persuasion and of mean-making—that is, stories as a basic way that humans make sense of and interpret the world. Our goal was to ground our analysis firmly in research, extracting relevant insights and best practices for narrative broadly, and then synthesize these insights and practices with a parallel line of research looking at narrative best practices in operational settings within the U.S. Department of Defense (DoD).

Method for Identifying Literature to Review

To increase rigor and relevance in our research, we followed a systematic literature review method (Tranfield, Denver, and Smart, 2003). We searched peer-reviewed literature to define best practices and the characteristics of narratives. Our search limited content to articles published in English from January 2000 through October 2019. We searched six databases: Academic Search Complete, PsycINFO, Scopus, Web of Science, the ProQuest Military Database, and the Defense Technical Information Center. Database selection was guided by the need to identify both scholarly and practitioner perspectives.

We developed a keyword strategy to broadly capture narrative literature. This included keywords specific to (1) communication and narratives, (2) linguistics, (3) psychology, (4) neuroscience, and (5) defense studies. Several test searches helped refine the final executed keywords. Each search combined a keyword from the communication keywords (10 in total) with a keyword from one of the other four categories (19 in total). Appendix B contains the keywords associated with each category.

The initial search produced 5,197 results; the results for each database are illustrated in Figure B.1 in Appendix B. After the removal of duplicate citations and 13 inaccessible Defense Technical Information Center articles, 2,174 unique articles remained for review. Because of the high volume of results, two reviewers constructed a labeling system to determine which articles were most crucial to review. Next, a reviewer read 2,174 article abstracts and binned them into one of four categories: crucial, highly recommend, low relevance, or delete (this last category was for articles that entered the dataset via relevant keywords but topically were irrelevant). This process identified 101 crucial articles, which were assigned to team members for review and additional data-extraction tasks. Moving into the full article review, we discovered 11 additional inaccessible or irrelevant articles, leaving 90 relevant, crucial articles.

We used a systematic data-extraction process to review each article. This included documenting several characteristics of each article: (1) type of source (peer-review journal article, organizational document, or other), (2) study methodology, (3) study target group (i.e., primary audience), (4) rhetor (i.e., the communicator in the study), (5) narrative definition, (6) components of a narrative discussed, (7) best practices for assessing narrative, (8) theoretical frameworks included in the article, (9) strategic communication best practices, and (10) an open field to capture any other significant findings from the article. Notes from the extraction were compiled into a master spreadsheet to identify key themes and findings from the crucial article sample.

We supplemented the data collected from the literature review with additional crucial sources provided by team members. Additionally, some sources were further identified during the full crucial article review process. These sources originated in the crucial articles and were gathered from the sample's reference list. Based on our literature review, we present in the rest of this chapter a summary of concepts relevant to narrative, including definitions, components and moderators of narrative, audience, and best practices.

Narrative Theory

In this first major section of findings for the chapter, we review elements of theory about and surrounding narrative, including definitions of narrative, the importance of narrative, and the purposes to which narratives are put.

Defining Narrative

Stories play a central role throughout our lives. Humans have "relied on stories to communicate and to archive . . . key events, histories, concepts, beliefs, and attitudes" (Haven, 2014, p. 3). Stories can be shared across time and cultures (Palmer, 2018). However, not all stories are remembered and disseminated broadly. Listeners typically default to filtering stories. Most stories capture an individual's awareness for a short time but are quickly forgotten (and with them, any lasting impact they could have left behind). However, practitioners and scholars have accumulated knowledge over time concerning what makes a story effective. While terms might differ across disciplines, storytelling constructs are present across industries, including journalism, marketing, psychology, sociology, and medicine, among others (Shaffer et al., 2018).

Narrative Versus Story

Narrative and *story* are not exact synonyms. A story is a sequence of events "involving actors and actions, grounded in desire (often stemming from conflict) and leading to an actual or projected resolution of that desire" (Corman, 2011, p. 1). Stories may be unique but often follow story forms, including typical actors, actions, and sequences (Corman, 2011, p. 1). Story forms use archetypes, standard characters, typical motives, and expected behaviors (Corman, 2011, p. 1). Building on these descriptions, some authors identify essential components to a story. For example, Haven describes eight essential components to stories: characters, traits, goals, motives, conflicts and problems, risk and danger, struggles, and details (Haven, 2014, p. 67). (These components are described in greater detail in a later section.) Using these defined principles, individuals can create their own stories to accomplish specified objectives.

A narrative, on the other hand, is a broader system of stories that unites themes, forms, and archetypes. The various stories within a narrative do not need to be identical, but they can be interpreted in a manner that creates a consistent system of information that shapes an audience's worldview (Corman, 2011, p. 1). Over time, narratives foster a shared connection of who we—as an organization or nation—are, what we do and why, and where and when we act (Nissen, 2013, p. 71). Not all narratives are intended to be persuasive or intentional (Shen et al., 2017). In fact, narratives can become powerful because they influence us without purpose or intent. Conversely, narratives can be developed to influence consumer beliefs and actions (Berger, Ha, and Chen, 2016). For example, a communicator can purposefully seek to convey information and understanding to an intended audience to influence how that audience perceives and thinks about a topic. Narratives are also used by dangerous actors, such as violent extremists, to exert control over others. Narratives used by ISIS (the Islamic State of Iraq and Syria), for example, can be compelling because they provide explanation of past events and contemporary conditions and incorporate ideas, texts, and interpretations from Islamic sources (Mahood and Rane, 2017).

Counternarratives can emerge to offer an alternative viewpoint to an existing narrative. Narratives can be effective for presenting value-discrepant information (i.e., information that conflicts with an individual's current beliefs) because narratives are less objectionable than blatantly contradictory appeals, including traditional factual messages and statistics (Shaffer et al., 2018). Narratives do not need to be based in truth, which is why harmful narratives can take root. Just like it doesn't make sense to say a poem is untrue or inaccurate, what is relevant about narrative "is that it strikes a chord in experience" (Maan, 2015, p. 3).

While there is a substantial amount of literature on stories and narratives, it is poorly organized (Corman, 2011, p. 1). Many authors have put forth various definitions of what constitutes a narrative or a story (Haven, 2014, p. 59–68; Braddock and Dillard, 2016, pp. 446–467). These terms are further conflated with additional constructs, such as "message" or "theme." Messages and themes are related to narratives but lack the total depth and complexity of a narrative.[1] Among the existing definitions, Daniel Bernardi's interpretation offers the most applicable definition for military applications. He defines stories as "having a sequence of events or actions, involving characters in specific settings, recounted for rhetorical purposes."[2] Stories are constituted by time (i.e., sequence of events), space (i.e., settings), representations (i.e., characters), and rhetoric (i.e., communication of an idea).[3] Narrative, on the other hand, is the system of interrelated stories. More specifically: "Narrative shares story elements, cultural references, and a rhetorical desire to resolve a conflict by structuring audience expectations and interpretations."[4]

The Importance of Stories and Narrative

The history of humanity is one of relying on stories to communicate and archive events. This reliance on stories is hardwired; simply put, our brains developed to "make sense, think, understand, and remember in specific story terms and elements" (Haven, 2014, p. 29). This is why narratives—systems of stories—are so powerful, because people automatically perceive the world with mental models that affect both what they perceive and how they interpret that incoming information (Haven, 2014, p. 2). Before incoming information enters our awareness, it has already been processed and shaped into story form (Haven, 2014, p. 29). This story takes the shape of what is known, or what makes sense to the individual, based on their lived experiences (Haven, 2014, p. 29). This filtering process is important to keep in mind, because we—as individuals—tend to filter out what we do not know, understand, or otherwise counters our experiences as we collect the stories that make up our narrative framework for understanding the world.

---

[1]   The Joint Staff J-7 has released guidance for CommSync, explicitly stating the differences between narratives, themes, and messages. Whereas a narrative focuses on "context, reason, and desired results [and] . . . enables understanding for external stakeholders," messages are "narrowly focused communications that support a specific theme . . . to create a specific effect." Also see Joint Doctrine Note 2-13, 2013.

[2]   This quote is from McGuinn, 2019; as a source, McGuinn cites Bernardi, Cheong, and Lundry, 2012, pp. 167–171.

[3]   McGuinn, 2019; McGuinn cites Bernardi, Cheong, and Lundry, 2012, pp. 167–171.

[4]   This quote is from McGuinn, 2019; as a source, McGuinn cites Bernardi, Cheong, and Lundry, 2012, pp. 167–171.

This reliance on narrative demonstrates the importance for approaching communication through a narrative lens. In comparison to other forms of communication, narratives present a more effective means of connecting an audience to a message and arousing an emotional response, influencing the audience's attitudes and behaviors, and helping reduce motivation for resistance and counter-arguing (Murrar and Brauer, 2009; van Laer, Feiereisen, and Visconti, 2019, p. 135). When an audience relates to a character or their struggles, the outcomes in a story can provide purpose or suggested courses of action (Paul, Colley, and Steckman, 2019, p. 81). As such, narratives can advance various agendas, including supporting pro-environmental behavior (Morris et al., 2019, p. 24), enlisting in the military (Paul, Colley, and Steckman, 2019, p. 81), generating support for terrorist organizations (Mahood and Rane, 2017, pp. 23–24; Paul, Colley, and Steckman, 2019, p. 81), or sharing health information (Shaffer et al., 2018, p. 430). Leveraging the human cognition, narrative can engage individuals in the same way that real-life experiences do. More specifically, the region of the brain that processes sights, sounds, tastes, and movement is the same region that responds to a compelling narrative (Haven, 2014, pp. 21–22).

The U.S. government, and more narrowly the military, can leverage the power of narrative to realize specific goals. This could be to support positive perceptions of one's organization or provide an alternative explanation to modify or replace the information in the audience's mental model that is unfavorable to one's organization (Hamby, Ecker, and Brinberg, 2019, pp. 242–242; Paul, Colley, and Steckman, 2019, p. 82). When seeking to defeat a hostile narrative, the sender must simultaneously promote a positive narrative in its place as, "the one thing that replaces [or modifies] a story-based belief . . . is a better story" (Seese and Haven, 2015, p. 34). Audiences *will* find a narrative or narrative frame to make sense of an event, one way or another. It is important however, to recognize that "narratives already exist" among an intended audience (Paul, Colley, and Steckman, 2019, p. 81). It is difficult to completely change or replace a narrative that an individual believes and uses to understand things that the person sees and experiences in the world. Instead, narratives should be developed in consideration of the narratives and beliefs already in use by a target population. In the following section, additional details are provided about the purpose of narratives, along with descriptions of several types of narrative.

Narrative Purpose

Narratives are an appealing tool for communicators because narrative form "make[s] it easy for your audiences to accurately understand, remember, and apply your material," with the goal of "affecting their attitudes and behavior" (Haven, 2014, p. 18). This could be directed at an internal audience for "internal organizational purposes" or to ensure that actions and communications are appropriately aligned or an external audience (Haven, 2014, p. 9; Paul, Colley, and Steckman, 2019). When successful, a narrative will influence an audience's attitudes, beliefs, or behavior to more closely align with the sender's. The focus of the narrative can vary. In marketing, a narrative may be used to persuade an audience to purchase a product. In medicine, "narratives, or 'illustrative examples of others' experiences,' are used to provide information, convey empathy, and change behavior" (Shaffer et al., 2018, p. 430). In a political context, narratives are used to influence how an audience thinks about a candidate, organization, or country and shape their attitudes and behaviors in support of it.

In a military context, command narratives might be necessary for several reasons. According to the Joint Staff J-7 focus paper on communication strategy and synchronization, one purpose of a compelling narrative is to help promote the legitimacy of the mission and "prevent the 'say-do' gap in which our actions and words conflict in the eyes of the audience" (Joint Staff J-7, Deployable Training Division, 2016, p. 1). For military staffs, narrative can "provide direction and guidance for communication on all levels," harmonizing words and deeds so the command sends a coherent message.[5]

Narratives can help guide both internal and external audiences and provide meaning for why the U.S. government is using the military. The U.S. government communicates with these various audiences, for different purposes, using different messages, themes, stories, and dissemination mechanisms. This communication, however, should not only occur at top levels. Personnel at all levels should serve as storytellers for the organization, whether it be a GCC or a private company. Encouraging personnel at all levels to serve as organizational storytellers helps to further disseminate a narrative and build relationships between the sender and the audience.

Internally, like a private business seeking to instill company values among its employees, the military may need to promote in-house coordination by providing a clear mission and message for individuals to follow. According to Paul, Colley, and Steckman (2019, p. 83),

> Such a narrative needs to fit with existing military and service-specific narrative frames [and when] done right, a mission narrative makes it easier for everyone to understand and remember mission objectives, and to understand their role in story terms that will lead to achievement of those objectives.

In addition, a clear narrative helps to unify individual efforts and ensure that the actions carried out are in alignment with the narrative being presented. Internal audiences could include U.S. citizens, organizations in the United States, the U.S. government, service members, and contractors, among others.

A second purpose of command narrative is to provide a positive or alternative narrative or framework for external audiences. For example, the military can provide a narrative for external audiences to interpret a U.S. operation. Specifically, a war narrative can establish desirability for a war policy through the development of relevant costs and benefits of the policy. The right narrative at the right time can help to pro-

---

[5]   Interview with personnel from the NATO Strategic Communications Centre of Excellence, March 9, 2020.

mote positive outcomes associated with an operation and encourage an audience to support (or least tolerate) the activity. If an alternative narrative is not available, and an external audience relies solely "on their own histories and experiences," existing and unchecked narratives "will support views and actions that are contrary to U.S. operations" (Paul, Colley, and Steckman, 2019, p. 83).

A third purpose of command narrative is to "compete with or undermine narratives at odds with mission objectives [as] many operating environments may contain narratives or narrative frames that do not support U.S. force presence or objectives" (Paul, Colley, and Steckman, 2019, p. 84). Counternarratives are considered an essential component of soft counterterrorism approaches, and, as adversaries continue to target the United States with influence campaigns, alternative narratives are needed to "deconstruct the militants' ideology and effectively challenge it with an aim to create an environment that counteracts the militants' narrative" (Iqbal, Zafar, and Mehmood, 2019, p. 147). Achieving this goal can be challenging, because adversaries may share cultural narratives with a population (Meier, 2016, p. 9). However, effectively developing counternarratives can have real and tangible positive outcomes, including a decreased duration of conflict and increased durability of postconflict arrangements (Meier, 2016, p. 27).

Narrative Types

The overall need and desired outcomes for communication can help direct the narrative purpose and type. For example, Shaffer et al. suggest five objectives for narratives: to inform, engage, model behavior, persuade, and provide comfort (Shaffer et al., 2018, p. 434). Narratives might focus on one or a combination of these objectives. Depending on the audience (i.e., internal versus external), and the objectives (i.e., inform, engage, model, persuade, comfort), different types of narratives might be more relevant to a specific situation.

A meta-analysis of narrative health studies describes three types of narratives: (1) process narrative, describing how an individual made a decision, (2) experience narrative, describing what it is like to undergo a certain experience, and (3) outcome narrative, describing the psychological or physical health outcomes associated with the activity (Woudstra and Suurmond, 2019, pp. 328–329). Each of these types focuses on a certain activity and/or behavior that the narrative might influence. Process narratives can help provide more context for decisionmakers. For example, process narratives of individuals receiving different types of cancer treatments can provide needed context for those newly diagnosed, who are faced with myriad options and combinations for possible care. Process narrative–related studies have found that addressing affective forecasting (i.e., individuals' predictions about how they will feel during an event) toward a health activity (such as screening) reduced perceived barriers to engaging in the activity (Woudstra and Suurmond, 2019, pp. 330–334). This could possibly translate over to non-health-related communication, speaking to the importance of addressing encountered struggles and barriers in a narrative as a way of getting ahead of how an audience may perceive barriers to behavior and attitude changes. Alternatively, experience narratives can help audiences understand what "something is really like" and provide a window into something that otherwise might remain hidden (Woudstra and Suurmond, 2019, pp. 330–334). Lastly, outcome narratives describe the consequences (positive or negative) of an event or decision.

Specific to the military setting, commands may use a different lens to view a narrative as either (1) a mission narrative, (2) an external narrative, or (3) a tactical-level narrative. Mission narratives communicate the commander's intent and the role troops play in achieving the desired end state and goal. The primary audience for this narrative are the troops and personnel charged with carrying out the narrative and operation's objectives. An external narrative, conversely, is inherently different because the audience holds different cultural assumptions from troops and associated personnel. An external narrative must effectively "incorporate an understanding of existing narratives, the frames that peoples use to view them, and how the command narrative nests under the U.S. or Coalition strategic narrative" to be successful (McGuinn, 2019, p. 21). A tactical-level narrative is more specific and scenario-focused, narrowing in on relevant actors and the populace within a joint operating area. A tactical-level narrative is vertically aligned, or nested under, the mission and operational narratives and relies on stories to support specific missions or engagements (and associated relevant actors) (McGuinn, 2019, p. 23).

Fiction and Nonfiction Stories

The stories that make up a narrative can be fiction, nonfiction, or a combination, as both fictional and nonfictional stories can create a persuasive narrative. Both fictional and nonfictional stories can involve realistic themes and resemble events that an audience may have experienced, giving the impression that the story events could happen to people (Berger, Ha, and Chen, 2016, p. 307). Stories that use real or fictional characters and events have their individual benefits and challenges. For stories with real characters and events, the audience can easily relate to and identify with these components, but the sender is limited to what actually occurred, because fabricating or changing things can cause the narrative to lose credibility. In contrast, fictional characters and stories can be made to say and do whatever the sender wishes, allowing the narrative to closely align to the characteristics of the target audience. However, the sender will "need a strong opening hook to the story" to help ensure that the audience does not perceive the narrative as unimportant or irrelevant to them (Haven, 2014, pp. 153–154). In addition, the sender should also be clear as to whether they are using fictional characters or stories to avoid loss of credibility with the intended audience (Haven, 2014, pp. 153–154).

Story Components

Stories play an essential role in narrative theory. Narratives comprise systems of stories that unite themes, forms, and archetypes. Studies define several different elements or components that make up effective stories within narrative. For example, all stories need characters, an overarching goal, obstacles and risks that the characters encounter in pursuing the goal, and a final resolution. Haven (2014, p. 67) identifies eight essential components for stories—characters, traits, goal, motives, conflicts and problems, risk and danger, struggles, and details—which we briefly describe in this section.

Characters and Traits

Every story needs characters, with descriptive traits, to help draw audiences in. Audiences will not care about a story if the characters are not compelling (Haven, 2014, p. 88). Research suggests that as audiences mentally "simulate the events that happen to a character, they may come to understand what it is like to experience the described events. As a result, their beliefs and attitudes may become more consistent with those of the character" (Braddock, 2015, p. 51). To help accomplish this, stories should be crafted specifically for a certain audience. In this sense, characters should reflect traits that are relevant to primary or secondary audiences. These traits may include the appearance, values, demography, or common experiences of characters (Shen et al, 2017, p. 166). Characters need to be a physical entity, possess a will (i.e., ability to think), be capable of acting in support of that will, and be able to communicate (Haven, 2014, p. 88).

These are concepts that extremist organizations, such as the Animal Liberation Front (ALF), use to their advantage. According to a study that assessed the ALF's narratives, prominent themes on the ALF's website highlight animals' cognitive and emotional capability to humanize them and their overall traits (Braddock, 2015, pp. 47–48). The organization uses storytelling, including character development, to communicate directly with audiences. In stories the ALF presents, animals share their experiences related to harsh punishments, mistreatment, and their ability to continue to express compassion and loyalty. Using anthropomorphism, ALF provides a way for humans to feel more deeply for the animals. By humanizing and developing the characters of the story, providing them admirable traits, and detailing the hardships they are enduring, the audience may be better able to connect with and identify the characters (in this case animals) in the story to create a more influential narrative.

Goals

What a character needs or wants to accomplish in a story constitutes a goal, which provides a purpose to the story (Haven, 2014, pp. 70–72). Goals should be physical and tangible, such as owning a home, because nontangible goals, such as happiness or peace, make it challenging for the audience to visualize and evaluate goal success. Audiences will interpret events and actions in a story against character goals, and so clearly articulating a goal ensures that audiences get the point of a story (Berger, Ha, and Chen, 2016, p. 307). Characters without goals can be hard to interpret and will be less likely to interest audiences (Haven, 2014, p. 70).

Motives

In addition to establishing a character's goal, the story should explain a character's motives (Haven, 2014, pp. 85–86). Motives are critical for two reasons. First, people need them to understand a story and will infer motives if they are not properly explained. When audiences try to infer a character's motives, they often create a more sinister or negative motive than what the storyteller intended. Second, motives create empathy and identification with the audience. Motives help the audience interpret the character and the story. Ideally, the motives will align with those of the audience, which can make the story more appealing.

Terrorist organizations convey organizational goals and motives in their stories to justify their actions and garner support. For example, justifying actions through the goal of upholding the Quran, ISIS and its potential recruits "are able to excuse, ignore or justify excesses and brutality for they may claim the ultimate goal of their struggle is establishing a just peace" (Mahood and Rane, 2017, pp. 26–27). Audiences that identify with someone's motives will be more likely to approve of a character and adopt their viewpoints, attitudes, and outlook (Haven, 2014, p. 113). ISIS can appeal to members and recruits by framing its motivation in context with Muslims living in non-Muslim, Western countries who feel victimized or discriminated against (Mahood and Rane, 2017, pp. 26–27).

Plot and Climax

The terms *plot* and *climax* are often referenced as elements in narratives; Haven does not use these terms in listing his eight essential story components but rather breaks them down into "conflicts and problems," "risk and danger," and "struggles." Conflicts and problems are what will block a character from achieving his or her goal (Haven, 2014, p. 113). After all, a story would not be very interesting without the characters overcoming some sort of challenge. Without overcoming a conflict or problem, audiences might see a story as pointless (Berger, Ha, and Chen, 2016, pp. 307–308). Creating understanding surrounding a conflict or problem can be helpful in instances where audiences currently have low awareness. For example, without interpreting climate change as a severe problem, it can be more difficult to persuade audiences to act (Morris et al., 2019, p. 31). Other, more tangible conflicts and problems may be easier to understand (e.g., losing a job, falling ill, struggling in school, immigrating to a new country). Some organizations, including the military, may hesitate to acknowledge the existence of conflicts or problems that can be detrimental (Haven, 2014, p. 74). Yet, audiences need this element to understand the experiences of the actor. Clearly linking the conflict or problems to the character, and to that character's goal, helps make a story worthwhile. Therefore, a good narrative must identify the goal that the protagonist wishes to achieve and the obstacles that must be overcome to achieve it (Davidson, 2017, p. 5).

Similarly, risk and danger contextualize the degree of difficulty and threats a character faces in achieving an intended goal—in other words, "the stakes involved—what stands to be lost or won" (Davidson, 2017, p. 5). While losing a job, falling ill, struggling in school, and immigrating to a new country all may constitute conflicts or problems for a character as they work to achieve a goal, it's not clear how likely these events are to happen or what exactly will happen to the character if one of them comes to fruition. For example, how likely is it that a character will fall ill (e.g., is the character young and healthy, or old and frail?), and what exactly will that mean for the character (e.g., they cannot work and/or provide for their family or continue to work on a project that affects national security)? Imparting risk perceptions for several health behaviors, including smoking tobacco or drinking alcohol, has been the goal of many public health education campaigns (Ooms, Jansen, and Hoeks, 2019).[6] Lower risk susceptibility can yield lower negative emotions and make a story less effective. For example, individuals may not curb alcohol consumption or related behaviors (e.g., drinking and driving) if the risks and dangers remain unclear.

The best stories also show characters struggling through their experiences. Struggles include characters expending a great effort to fight for what they want, pulling through their conflicts and problems, and facing risks and dangers to ultimately achieve their goals (Haven, 2014, p. 113). Rather than treating struggles as an indication of weakness, storytellers should "treat struggle as a strength or virtue" that leads to the eventual outcome of whether the character was successful or not (Haven, 2014, p. 83; Berger, Ha, and Chen, 2016, p. 303). Further, struggling can help audiences become more empathetic and supportive. The theme of struggle and suffering can be found in many types of life stories, which can help audiences understand and admire a protagonist.

Details

Details paint a picture and allow the audience to better imagine the story. Details about characters and their surroundings bring them to life. Communicators should work to include elements that speak to each of the five senses in order to help audiences envision a story to its fullest potential. A single word can add detail to a story and can contribute to how a character is experiencing the world (Forman, 2013, p. 36). Further, adding details ensures that an audience will see the story in its intended light—audiences will create their own details if they are not provided (Palmer, 2018, p. 504). Details can be challenging for novice writers. Artful descriptions can differ from what many writers might be used to in other forms of communication, and, technically,

---

[6] See also Kubacki et al., 2015, pp. 2214–2222.

stories could be written without details. However, memorable visualization sets some stories apart from others and makes them more likely to be committed to memory (Haven, 2014, p. 113).

Story Moderators

The eight essential components provide the key mix of ingredients to create effective stories. However, additional literature provides insight into several moderators that may further explain why some stories are better than others. These include causality and temporality, coherency, audience connection and proof, and plausibility. These moderators address aspects of the structure and formatting of a narrative.

Causality and Temporality

Stories should be composed of causally and thematically related events that follow a temporal order (Shaffer et al., 2018, p. 438). These features enable the audience to follow how elements depicted in a narrative link together. Causal, temporal formats further conform with how people commonly recall information. People recall narratives from memory in a linear, cause-effect format even if the narratives were not originally presented that way (Hamby, Ecker, and Brinberg, 2019, p. 243). If a piece of this story chain is eliminated, it creates a gap and hinders the coherency of the story (Hamby, Ecker, and Brinberg, 2019, p. 241). Stories that do not follow this convention are more challenging to commit to memory and subject to revision as time passes.

Temporal message framing can provide context as to when a character experiences risks and/or danger. One study found that participants exposed to positive, short-term messages reported behavior changes related to alcohol consumption (Gerend and Cullen, 2008). This aligns with what other literature suggests, which is that individuals tend to discount the future (Estle et al., 2006). However, temporal variance is best approached by placing it in the context of the topic at hand along with the characteristics of an audience. For instance, one study found that low-autonomy individuals were most likely to be affected when negative, short-term consequences were conveyed (Churchill et al., 2016, pp. 272–733). In this sense, attending to the construct of time and how it intersects with existing story constructs offers another opportunity for impact.

Coherency

Coherence refers to a narrative's need to maintain a structured internal consistency (Morris et al., 2019, p. 21), as well as unity and integration with the audience's understanding of stories and events (Hamby, Ecker, and Brinberg, 2019, pp. 242; Mahood and Rane, 2017, p. 17). As noted previously, incoming information is first processed and shaped into story form by the receiver (Haven, 2014, p. 29).

A narrative should align with the receiver's existing repository of stories and memories to improve narrative coherence. However, if a narrative lacks instant coherence, there should be an alternative causal explanation to fill in gaps and improve overall coherence (Hamby, Ecker, and Brinberg, 2019, pp. 242). The lack of an explanation can create a coherency gap that can be filled in—either unintentionally by the receiver or intentionally by actors seeking to manipulate or influence an audience. Thus, avoiding coherency gaps is important when either developing a new narrative or seeking to modify a current narrative among an audience.

Recent studies in marketing and consumer psychology illustrate that "people process stories in ways that are distinct from the processing of argument-based messages" (Hamby, Ecker, and Brinberg, 2019, p. 240). With stories, people seek to link elements together to create a cause-effect chain; a missing element creates a gap and hinders the story's coherence. As such, "past work hypothesized [and this study validated] that (mis)information embedded in a story influences readers' reasoning and beliefs because people implicitly prefer a complete (but potentially incorrect) model of an event over a correct but incomplete model; that is, discrediting causal information embedded in an accepted cause-effect model can create an undesirable state of incompletion" (Hamby, Ecker, and Brinberg, 2019, p. 241). This helps to explain why some people continue to rely on information even after it is shown to be inaccurate, such as the danger of vaccines. To reduce the likelihood of false information continuing to influence an audience, the sender must provide an alternative and clear explanation for an event or issue that fills the gap. These points about coherency demonstrate the importance of communicating information in a way that is relatable to the audience and will allow the audience to pull from their knowledge and prior experiences (Haven, 2014, pp. 134–136).

Character Identification and Proof

Building on coherence, there must also be alignment between characters and the intended audience. Audiences that identify with characters will find a narrative more powerful, and audience identification with characters can give the narrative more credibility (Woudstra and Suurmond, 2019, p. 334). According to the Narrative Paradigm, a communication theory conceptualized by 20th century communications scholar Walter Fisher, people are persuaded to make decisions based two aspects of stories (Morris et al., 2019, p. 21). The first is coherence (i.e., internal consistency of characters and plot), and the second is fidelity. Fidelity includes the external consistency of the narrative's values with the audience (Morris et al., 2019, p. 21). These connections to the audience can be based on a variety of factors, including appearance, values, demography, or common experiences, which help to create a sense of familiarity with the narrative (Shen et al., 2107, p. 166; Paul, Colley, and Steckman, 2019, p. 82).

Each audience has their own beliefs and perspectives, which affect how they interpret incoming actions and words (Joint Staff J-7, Deployable Training Division,

2016, p. 2). As such, proof, or evidence that supports the credibility of the narrative and sender, can vary depending the local context and medium. In the United States, for example, "the facts in stories presented by television news anchors are accorded high degrees of credibility and generally accepted as strong proof" (Paul, Colley, and Steckman, 2019, p. 82). However, in some countries, media are not afforded as much credibility because of their ties to government. In these countries, a story repeated by a friend might count as stronger proof despite lacking strong evidence (Paul, Colley, and Steckman, 2019, p. 82).

Plausibility

In addition to these factors, compelling narratives also have plausibility. This refers to whether the audience perceives the narrative as realistic or plausible, which can help activate empathic responses in the audience (Shen et al., 2017, p. 167). Ensuring that a narrative is plausible in the eyes of the audiences helps to address instances of the audience potentially dismissing a story as far-fetched or unlikely (Berger, Ha, and Chen, 2016, p. 307).

Audience Response

The literature typically uses emotional reactions and narrative transportation—the degree to which an audience is absorbed into a narrative—to gauge influence. Both positive and negative emotions can demonstrate the impact of a narrative and the specific stories within a narrative, in addition to transportation, or the degree to which an audience is absorbed or immersed into a narrative.

Emotional Response

Stories can be powerful because they can stir emotions from receivers. According to Clausewitz, "Truth in itself is rarely sufficient to make men act. . . . The most powerful springs of action in men lie in his emotions" (quoted in Meier, 2016, p. 31). Information and facts alone may not influence thoughts or reactions. Further, evoking emotions is attention-grabbing. This helps the sender control or mediate how an audience responds to the story. Evoking emotions also ensures that the story is memorable enough to stay with the receiver once the storytelling ends.

Emotions arise as an audience processes the implications from events and considers the outcome to attaining a goal (Dillard and Nabi, 2006, p. S124). For example, is being fired from a job devastating to a character (e.g., it was the character's dream job), or might it bring the character closer to self-actualization (e.g., it was distracting the character from what they were destined to do)? In the case of the former, we might feel sad or angry. In the case of the latter, we might feel relief and excitement. Of course, storytellers need to anticipate the emotional reactions of their audiences and ensure that character identification will elicit the intended reactions. The ability of a given story to cause a strong emotional reaction among the audience serves as an indicator of how effective a narrative may be, whether the focus is on generating support for a military operation, supporting pro-environmental behavior, or sharing health information.

Haven finds that stories described by the audience as *powerful, meaningful, impactful,* or *memorable* (all surrogates for *influential*) are the stories that elicit emotional reaction (Haven, 2014, p. 29). A study comparing the impacts of climate change information found that story structure was more effective in encouraging action-taking behavior than were factual appeals (Morris et al., 2019, p. 31). In the study, storytelling physiologically triggered responses in participants, such as changes in heart rate (Morris et al., 2019, p. 31). Emotions also help individuals remember stories long after they're told. Haven describes the importance of residual emotions, which include how someone feels after the story ends (Haven, 2014, p. 29). Great stories leave their audiences emotional once the story has reached a resolution. While audiences may still enjoy stories without residual emotion, such stories are much less likely to truly persuade or change people (Haven, 2014, p. 29). Haven concludes that "there exists a direct correlation between the strength of the ending emotional state of the audience and the magnitude of the influence that story can carry" (Haven, 2014, p. 104).

Positive Versus Negative Endings

Both positive and negative ending emotions can be effective and influential. However, encouraging a positive or negative response may lead to different types of outcomes. Storytellers should determine what type of effect they hope to achieve to select the appropriate approach. Positive emotional endings "do not motivate immediate action nearly as well as negative endings," but positive story resolutions work on a much longer time scale that can shape future behavior (Haven, 2014, pp. 127–129). In other words, the more negative an emotional ending is, "the greater the tendency for that audience member to take immediate action (e.g., donate, protest, write letters, boycott, riot, etc.)" (Haven, 2014, p. 130).

An example of this distinction can be seen in footage of an Israeli excursion into the Gaza Strip in October 2010 (Haven, 2014, p. 131). The footage depicts a Palestinian father cradling his young son, who is hysterical as they hide behind a wall with gunfire in the background. Next, it shows the father behind the wall cradling his dead son in his lap (Haven, 2014, p. 131). Eventually, the footage was determined to be fake, but the video had already sparked a strong emotional reaction from its target audience. The footage sparked riots and "became the most effective recruiting and fundraising piece for the most extremist factions of the Palestinian and Arab communities" (Haven, 2014, p. 131). With regard to encouraging long-term activity, such as pro-environmental behavior, some research suggests that communicators should consider utilizing negative valence endings in narratives to help prompt "the perceived need for action" (Morris et al., 2019, p. 32). Additional insight about gain and loss framing, a related topic, is addressed in the best practices section.

Narrative Transportation

When stories are truly compelling, people can feel like they are transported into the action. Whether stories are told orally, through film, or through other media, people can feel like they are "entering the world of the story" (Hester and Schleifer, 2016, p. 109). This process of being absorbed, or "losing oneself in a story," allows the audience to vicariously experience the events depicted in the story. Stimulating transportation effects is a critical aspect of narrative success (Hester and Schleifer, 2016, p. 10). When an audience is fully immersed in a story, they can start to process information from the perspective of one or more of the characters involved (Shaffer et al., 2018, p. 437). Substantial research suggests that the effects of this phenomenon include reduced counterarguing to messages, increased connection with characters, and the greater likelihood of an audience (Morris et al., 2019, pp. 21–22), as Green and Clark (2013, p. 477) put it, "to change their real-world beliefs and behaviors to match those implied by the story." When narrative generates stronger feelings of transportation, there is greater opportunity to align the audience's beliefs and behaviors with that of the narrative. According to one study (Shaffer et al., 2018), the immersion process is a hierarchical continuum that the audience can move through, as noted in Figure 2.1. The first level ensures an audience's interest in a narrative (Shaffer et al., 2018, p. 436). The second level requires active participation by the audience, such as character identification and taking their perspective (Shaffer et al., 2018, pp. 436–437). The third level is full immersion into a narrative, where people begin to process information from the character's perspective (Shaffer et al., 2018, pp. 436–437).

One prominent study (Green and Clark, 2013) involved several experiments to determine the effects of narrative transportation on augmenting beliefs consistent with the story. The authors identified several components of transportation by an audience, including emotional reactions, descriptive mental imagery, "and a loss of access to real-world information" (Green and Brock, 2000, p. 703). In a series of experiments, the authors concluded that when audience members were highly transported into a story, their beliefs were more aligned with the story's conclusions, that the audience held

Figure 2.1
Narrative Immersion Process

SOURCE: Adapted from Shaffer et al., 2018, pp. 436–437.

"more positive evaluations with the story protagonists," and that this was unaltered by whether a story was fictional or not (Green and Brock, 2000, p. 707). In contrast, participants that experienced less transportation reported "reduced story-consistent beliefs as well as reduced positivity of character evaluations" (Green and Brock, 2000, p. 717).

Character Identification and Self-Activation

Characters can have a profound influence on how an audience perceives story. A compelling story can lead an audience to subconsciously identify with actors and the struggles they face. When people relate to a character or their struggles, the outcomes in a story can provide purpose or suggested courses of action (Paul, Colley, and Steckman, 2019, p. 81). For example, consider military recruiting commercials that depict a young person encountering personal challenges and wanting to pursue the noble goal of protecting people. Potential recruits who identify with the character in the commercial may be more encouraged to join (Paul, Colley, and Steckman, 2019, p. 81). Adversaries employ the same approach, including ISIS via the organization's *Dabiq* magazine. *Dabiq* articles describe the need to repel infidel oppressors from Muslim lands, employing characters, struggles, and goals that resonate with their intended audience (Mahood and Rane, 2017, pp. 23–24; Paul, Colley, and Steckman, 2019, p. 81). Character identification may also spur self-activation. Self-activation occurs when a narrative depicts experiences that are directly relatable to an audience, allowing them to perceive the actions as happening to themselves, rather than perceiving events from the character's perspective. (Zhou and Shapiro, 2017, p. 1298). Self-activation gives the narrative more credibility and evoking thoughts consistent with the narrative's message (Zhou and Shapiro, 2017, p. 1298). People can become acutely aware of self-related consciousness, including knowledge of a specific issue and/or personal associations related to that issue (Zhou and Shapiro, 2017, p. 1298). Zhou and Shapiro (2017) found that this perspective was useful for narratives targeting people engaging in binge drinking. Self-activation can be useful for reducing resistant behaviors such as counter-arguing or perceiving the narrative as unrealistic (Zhou and Shapiro, 2017, p. 1306).

Audience Understanding

Clearly, to reach one or more audiences, communicators need to know something about the people they intend to influence. Ideally, communicators will start thinking about who constitutes their audience at the beginning of narrative development. Communicators should treat audiences as both authors and recipients of any narrative (Zalman, 2010, p. 6). The historical experiences and cultural outlook of audiences constitutes part of their worldview, and communicators need to anticipate how stories will be interpreted in their own terms (Zalman, 2010, p. 6; Paul, Colley, and Steckman, 2019, p. 81).

Audience Analysis and Segmentation

Commanders and staff need to understand relevant audiences within their operational area. Audience analysis—considering each audience's own beliefs and perspectives, which inform how they perceive the U.S. military—can inform narrative development (Joint Staff J-7, Deployable Training Division, 2016, p. 2). Remember, too, that part of the context that shapes individuals' worldviews comes from culture and culturally prevalent stories or master narratives (Hochman and Spector-Mersel, 2020); identifying these should be central in audience analysis. Audience analysis is a formative step, based on primary or secondary research, that helps to uncover audience characteristics and needs (Mah, Tam, and Deshpande, 2008). In some cases, elements of a communication initiative may even be pretested with a sample of the intended audience (Mah, Tam, and Deshpande, 2008). Primary research could include interviews, focus groups, surveys, or other ways of observing audiences. Secondary research could include any review of existing information, such as previously collected data, literature reviews, or other available articles or sources. Secondary research is more economical, but the applicability of the information to a new project can be limited (Mah, Tam, and Deshpande, 2008).

Audience analysis might be challenging with some audiences, particularly when the audience might not trust the communicator (e.g., the U.S. government). However, methods such as observations, content analysis, or analysis of previous research can still be helpful during story development. One study, focused on minimizing harm from alcohol consumption, included observational studies in certain bars (Kubacki et al., 2015). Content analysis continues to be popular for following trends and patterns related to public sentiment, attitudes, and beliefs (Lam and Hannah, 2016). Traditional and new media provide many potential corpora that can be analyzed manually or with the assistance of natural language processing. Additionally, using data archives or other existing sources can be helpful for providing historical context and background on audiences (Noar, 2006). Regardless of the strategy, communication campaigns are more successful when developers engage in formative research to understand their audiences.

Audience Segmentation

Using insight gathered from audience analysis, communicators can create defined audience segments. Audience segmentation is the process of identifying unique groups of individuals and the specific needs and motives that distinguish one group from another (Kubacki et al., 2015). Segments can consist of groups that are of direct interest (i.e., a primary target audience) or a secondary segment (i.e., peripheral groups). Peripheral groups are those who may influence those in the primary audience segment (Key and Czaplewski, 2017). In the marketing world, segmentation can help pinpoint the right "marketing mix" (e.g., price, product, placement, and promotion) to satisfy different consumer needs (Kubacki et al., 2015). Segmentation requires an understanding of something that a group of people have in common, which can range from their job type, demographic characteristics, desires, issues or problems, motives, beliefs, experiences, or readiness to change (Mah, Tam, and Deshpande, 2008; Haven, 2014, p. 140).

Communicators can decide to target a segment based on its members' readiness to change. For example, segments could include those who are already engaging in some but not all aspects of a desired behavior (Cheng, Woon, and Lynes, 2011). Conversely, many behavior change efforts may opt to target audiences who are resistant to any number of behaviors (e.g., handwashing, substance abuse, radicalization). Behavior change theories can help point to ways that stories can frame conflicts and problems and how characters respond to those difficulties. For example, the precaution adoption process model seeks to understand prevention behavior and how others can influence an individual's perceived vulnerability and preventive responses (Weinstein, 1988). The theory posits that when an individual is aware of a problem in their community, know people are concerned about the problem, and know someone that has engaged in protective actions related to the problem, they are more likely to act themselves (Hinyard and Kreuter, 2007). Story characters can help people move closer to taking desired actions by acknowledging their current cognitive-stage of change.

In many cases, segments might simply be based on gender, age, race/ethnicity, education, geographic location, or language. However, more in-depth audience analysis may reveal more complex psychosocial and theoretical characteristics to guide audience segmentation. Communicators can use single or multiple layers of characteristics to define ideal segments (Slater, 1996). The more sophisticated segmentation becomes, the more opportunities communicators will have to create stories that resonate with audiences. Additionally, increasingly defined segments can maximize resources. Smaller segments can help cut out unnecessary outreach (e.g., media channels such as cable television) if communicators know how audiences prefer to engage in content (e.g., purely online or through interpersonal sources).

One program, funded by the European Union and tasked with developing guidance for managing communication during a future public health emergency, notes that audience analysis is essential for appropriately targeting different audiences (French, 2016, pp. 138–142). This is especially true in Europe, where country-level and even regional segmentation is needed (French, 2016, pp. 139–141). Another antismoking campaign in Europe used segmentation at the country level, recognizing that cultural attitudes toward smoking vary across European Union member states (Gianfranco et al., 2010, pp. 18–23), and that attitudes toward quitting smoking (i.e., strong desire to quit versus those with little to no intention to quit) had to inform the campaign (Gianfranco et al., 2010, p. 9). Identifying the most important variables of interest, including cultural variations among countries and attitudes toward smoking, helped create ideal segments for the antismoking media campaign. A failure to engage in audience analysis and segmentation can become a fundamental weakness in communication initiatives (French, 2016, pp. 138–142). This same principle holds true for those developing stories and narratives.

Best Practices for Improving Communication

Additional best practices in the literature provide perspective for improving narratives and storytelling. Targeting, tailoring, and framing are persuasion techniques that can be used in conjunction with narrative theory.

Targeting

The data generated through audience analysis and segmentation should be used to target communication to the appropriate audiences. Message targeting specifically "customizes messages to shared characteristics of population subgroups, such as lifestyle factors like recent college graduates in emerging careers in small cities or physically active retirees living in the suburbs" (Schmid et al., 2008, p. 32). This approach assumes that group members who share something in common will be influenced by the same message (Schmid et al., 2008, p. 32). Communicators can engage in targeting based on any defined audience characteristics (as discussed previously). For example, health communication efforts have taken advantage of targeting high-sensation seekers, which is a personality trait associated with the need for "novel, complex, ambiguous, and emotionally intense stimuli and the willingness to take social risks" to obtain stimulation (Stephenson and Palmgreen, 2001, p. 50). This information provides direction for message design features. For example, high-sensation seekers are more likely to prefer "high sensation value messages," which elicit emotional and physiological responses (Stephenson and Palmgreen, 2001, p. 49).

Targeting also typically directs channel selection, based on the needs of the target audience. Channels are various types of media mechanisms, such outdoor, print, radio, mail, social media, television, and websites, among others. No matter how persuasive a message is, selecting the wrong channel can ensure that the content never reaches the intended audience. Traditionally, health campaigns relied on television, radio, and print channels (Noar, 2006). For these traditional channels, national surveys and polls from Pew Research, Gallup, and Nielsen Media Research measure media channel preferences of consumers over time. However, since the mid-2000s, new opportunities to reach audiences online have changed channel targeting options. Online behavioral advertising can quickly assess user data, such as websites visited, articles read, and videos watched, in addition to anything typed into a search engine, in order to target advertisements (Boerman, Kruikemeier, and Borgesius, 2017). Further, social media outlets, such as Facebook, Instagram, LinkedIn, and Twitter, offer targeting advertising options to align paid content with relevant audiences. Targeting has become a key strategy in political campaigns, where campaign messages are able to "microtarget"

Figure 2.2
An Example of Audience Targeting

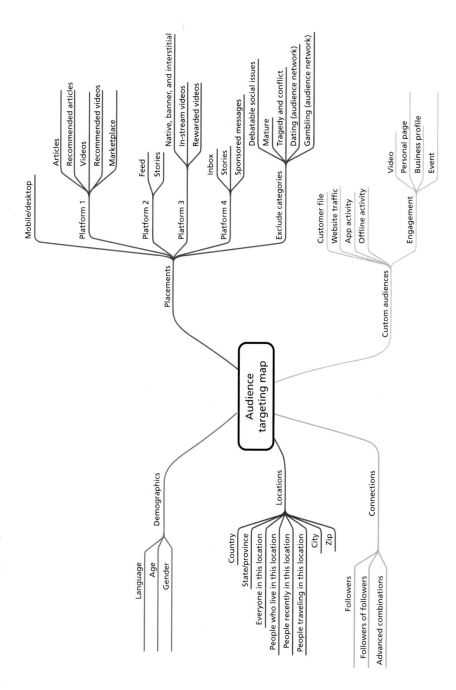

specific messages to increasingly narrower audiences (Ballard, Hillygus, and Konitzer, 2016, pp. 414–419). Figure 2.2 provides an example of how social media companies might target certain groups based on common characteristics, such as demographics, interests, or behaviors.

Tailoring

Tailoring goes beyond reaching audiences in large segments. Message tailoring involves a process that crafts parts of a message at an individual level. The most basic form of tailoring is personalization, "which involves identifying and incorporating into messages characteristics of an individual such as the person's name or age" (Schmid et al., 2008, pp. 32–37). However, tailoring can also employ more sophisticated approaches based on psychographic factors. For example, search engine results tailor content to each user, social media advertisements offer specific product suggestions, retailers use shopping history for coupons and sale information, and companies such as Netflix and Amazon tailor movies and other products on their platforms to each user (Noar, Harrington, and Aldrich, 2009, pp. 75–76). Ultimately, tailoring is persuasive because individuals perceive the message to be personally relevant to them.

Tailoring can be challenging, largely because of the expense associated with enhanced customization. However, it can be important when there is a range of knowledge, attitudes, or beliefs within audience segments (Noar, Harrington, and Aldrich, 2009, pp. 74–75). Other pitfalls could be that individuals can find too much tailoring "creepy or inappropriate" (Boerman, Kruikemeier, and Borgesius, 2017). In other words, if people feel that a message knows too much about them, they may feel like their privacy has been violated and disengage with the content. There is a tipping point associated with both targeting and tailoring that communicators should consider carefully. Frequently, targeting and tailoring terms are used interchangeably. However, targeting and tailoring are distinct tactics that can be used independently or simultaneously. Both activities help customize content in a way that is more relevant for intended audiences, which is more likely to result in audience effects (de Graaf, 2014). Figure 2.3 shows the conceptual link between narrative and the specific, targeted messages directed at various audiences.

Framing

The concept of how a message is framed has received a great deal of attention in the strategic communications field, including marketing and political campaigns, as one element that can help "lead to cognitive and behavioral changes" (Kim, 2014, p. 1). Message framing is how an issue is phrased, and its effect, "is related to how mental representations of the choice [issue], partly developed by message framing, influence cognitive or behavioral responses, including judgment and choice" (Kim, 2014, pp. 1–2). The origins of positively and negatively framing messages are linked to prospect theory (Kahneman and Tversky, 1979), which suggests that individuals "avoid risk for gains

Figure 2.3
Tailoring Messages

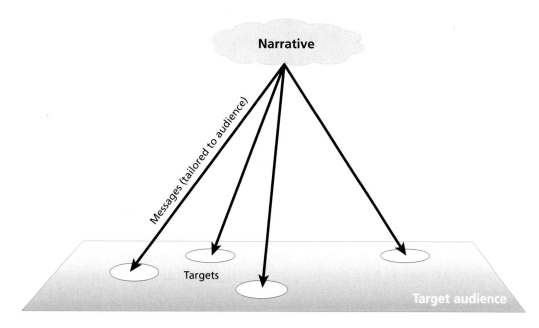

but we will take risks to avoid loss" (Sallis, Harper, and Sanders, 2018, p. 513). Health messages, for example, can be framed to showcase the benefits of doing something (i.e., gain-framed) or the consequences of failing to do something (i.e., loss-framed) (Gallagher and Updegraff, 2012, pp. 101–116). In health interventions, gain-framed messages are more persuasive for prevention behaviors (i.e., using sunscreen), but loss-framed messages are better suited for promoting detection behaviors (i.e., medical screening) because detection behaviors may be perceived as risky (Sallis, Harper, and Sanders, 2018, p. 513). However, additional elements can influence the outcome of message framing, such as the target of the message (i.e., person receiving the message or another contact, such as a friend or family member) and/or the temporal focus.

Putting framing theory into practice, Cheng, Woon, and Lynes (2011) presented participants with a hypothetical treatment program framed in terms of losses. Patients were more likely to prefer "a program in which 66% of all 600 patients will die over one in which there is a 100% chance that 400 of the 600 patients will die." Comparatively, when patients were presented with programs framed in gains, they were more likely to choose a program that has a "100% chance that 200 of the 600 patients are saved over one in which there is a 33% chance that all 600 patients are saved" (Cheng, Woon, and Lynes, 2011). As discussed above, patients opted for the riskier choice in the loss-framed condition and avoided risks in the gain-framed condition (Tversky and Kahneman, 1981, pp. 453–458). Figure 2.4 shows how variable framing can be,

Figure 2.4
Narrative Framing

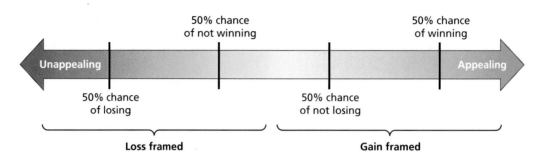

depicting multiple ways of choosing to frame messages with essentially the same factual content.

Temporal message framing may also influence the impact of gain- or loss-related frames. Temporal framing situates a risk occurring along a time continuum (e.g., short, medium, or far into the future). Of course, it should be no surprise that short- and long-term risks may be meaningful in different ways to people. Short-term risks might be more likely to prompt action now, compared with something that could be a far-off threat. For example, one study showed that undergraduate students enrolled in an alcohol consumption study were more likely to report consuming less alcohol if they were enrolled in the gain-frame/short-term consequences condition (Gerend and Cullen, 2008, pp. 1167–1173). Another study exploring the interaction between framing, temporal focus, and perceived individual autonomy found a significant three-way interaction between the constructs (Churchill et al., 2016, pp. 727–733). Thus, communicators should consider framing in crafting narrative. For example, communicators might consider what the goal of a narrative is (e.g., gain- or loss-related) and use that framing in crafting narratives.

Assessment

Evaluating the impact of narrative requires time and resources. Conducting evaluation activities can be challenging, but if communicators are not assessing narrative efforts, it is impossible to know whether the activities are making any difference. Evaluation or assessment is often used to describe "different things to different users in different contexts," but in the case of narrative evaluation, guidance typically conceptualizes evaluation to be about "measuring the performance or effectiveness of specific tasks, actions, events, or programs" (Paul and Matthews, 2018, p. 2). In other words, communicators want to know whether their future, in-progress, or completed efforts are making a difference.

Figure 2.5
Communication Assessment Types

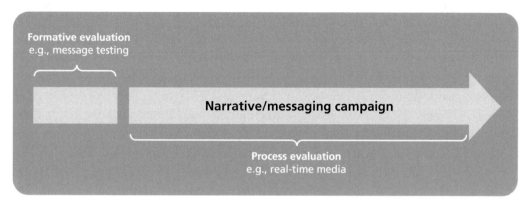

Summative evaluation

Communication assessment is broken down into three types: formative, process, and summative evaluation. Figure 2.5 shows the relationship between these types.

Formative evaluation includes audience analysis, as discussed previously, but also could include message testing. Pretesting messages with a small subset of audience members can help ensure that the messages are appropriate and effective (Noar, 2006). Process evaluation occurs while something is in progress. For example, communicators might entail collecting media data in real time to ensure that the messages are being delivered as intended (i.e., are disseminated on the right channels to the right audiences). Finally, summative evaluation seeks to understand the impact of communication on audiences. This includes identifying the total effects of the communication on audiences and calculating the change in specified outcomes (e.g., knowledge, attitudes, behaviors).

Using Theory to Guide Assessment

The theoretical approach for assessment should be determined at the beginning of any communication intervention. Using theory can help clarify the goals of communication, "explicitly connect planned activities to those goals, and support the assessment process" (Paul, 2017, p. 9). Theory can aid in the conceptualization of constructs and suggest certain determinants that guide campaign messages (Noar, 2006). A variety of behavior change theories are used in literature. Examples include social norms (Sallis, Harper, and Sanders, 2018, p. 513), the elaboration likelihood model, social cognitive theory, the precaution adoption process model, the theory of reasoned action (Hinyard and Kreuter, 2007, p. 783), the theory of planned behavior, the transtheoretical or stages of change model, social exchange theory, and value-belief-norm theory, among others (Cheng, Woon, and Lynes, 2011, pp. 49–50). Without theory, understanding which elements of a campaign or narrative mediated communication success can be difficult.

One study used the theory of planned behavior in Scotland to reduce speeding. Using three of the theory's main predictors attitudes, subjective norms, and perceived behavioral control, the study was able to change attitudes and affective beliefs surrounding speeding (Stead et al., 2005, pp. 36–50). Another study in the United Kingdom (Sallis, Harper, and Sanders, 2018) used social norms—the rules or standards understood by a group that constrain social behavior—to develop an organ donation campaign. The theory assumes that knowing what others do in a situation can guide the perceptions of what decisionmakers should do. For example, in the organ donation campaign, some web users received a prompt that said, "Every day thousands of people who see this page decide to register [as an organ donor]" (Sallis, Harper, and Sanders, 2018, p. 513). The study presented several other survey conditions and determined that social norm messages inspired more individuals to register as donors when compared with control messages.

Evaluation Design

There are many ways to design an evaluation approach. In an empirical setting, researchers can conduct studies in a lab that are able to pinpoint effects associated with specific variables (e.g., different message content, style, form, among other possible variables). [7] However, what these studies gain in terms of internal validity (or their ability to establish a true cause-and-effect relationship) they may lack in ecological validity (or if the theory/variables of interest would hold true in a real-world setting). Command narrative, for example, could be difficult to test in a lab, and thus likely needs alternative forms of evaluation design.

One approach is to specify and track various indicators, which could come from sources such as key leader engagements, news reports, or other secondary data (Paul, 2017, p. 53). Indicators are appealing because they are readily available and do not require sophisticated research designs. A considerable downfall, though, is that changes in indicators over time cannot be assumed to be caused by a specific communication initiative. Ultimately, the best evaluations use many measures and methods to triangulate results. Triangulation means that multiple intersecting measures are used to provide greater confidence than a single measure alone (Paul, 2017, p. 14). This type of approach combines the benefits of potentially analyzing print news, social media, and interviewing members of the intended audience to search for the presence of implemented stories or narrative.

Some studies with considerable resources can conduct large scale, multi-country evaluations. For example, one study included more than 24,000 respondents across 25 EU member states (Walsh et al., 2010, p. 11). The study's intervention included three

---

[7]   See, for example, Jin, Phua, and Lee, 2015, pp. 6–17; de Graaf, 2014.

commercials, which were aired twice a year on television (i.e., MTV, Eurosport, and Euronews) (Walsh et al., 2010, p. 10). This design allowed the researchers to examine perceptions about smoking across countries, in addition to other individual character-istics of the EU respondents (e.g., age, gender, education). The design also provided insights into key outcomes, such as awareness and recall. Evaluation designs can range from simple to elaborate and vary based on available resources and the needs of the project.

The Joint Staff J-7's *Commander's Handbook for Assessment Planning and Execution* (2011, p. IV-3) offers the following six-step process for planning and conducting assessments:

1. Gather tools and assessment data.
2. Understand current and required conditions.
3. Develop assessment measures and potential indicators.
4. Develop the collection plan.
5. Assign responsibilities for conducting analysis and generating recommendations.
6. Identify feedback mechanisms.

Outcome Measures

The outcome measures should be informed by the guiding theory. However, Table 2.1 contains an overview of some of the most common communication outcome measures used to measure communication effects.

The list of measures in Table 2.1 is not exhaustive. The included examples are well documented in literature and are generally well vetted when it comes to scale development. Evaluators should track the constructs relevant to specified goals and minimize unnecessary data collection requests. Again, the Joint Staff J-7's *Commander's Handbook for Assessment Planning and Execution* offers proves helpful by offering a framework for considering the various types of measures, shown in Figure 2.6.

Using Stories

In this literature review, we have emphasized the powerful impact that stories wield over us. Our brains are developed to "make sense, think, understand, and remember in specific story terms and elements" (Haven, 2014, p. 3). Before incoming information enters our awareness, it has already been processed and shaped into story form (Haven, 2014, p. 29). This story takes the shape of what is known, or what makes sense to the individual, based on their lived experiences (Haven, 2014, p. 29). This filtering process is important to keep in mind, because humans tend to filter out what they do not know, understand, or otherwise counters their experiences.

Table 2.1
Common Outcome Measures

Construct	Definition
Attitudes	The way a person feels about performing or not performing a behavior (Ajzen, 2011, p. 1117)
Beliefs	An understanding about the state of the world or the facts as we see them
Behavior	Objective information about an individuals' actions
Emotion	Positive and negative affective states (Morris et al., 2019, pp. 31–32)
Engagement	"Emotionally laden attention" (Haven, 2014, p. 18)
Exposure	Acknowledging seeing, hearing, or reading a message
Intensions	Reported, but not actual, behaviors
Identification	Holding something in common with a story character (Haven, 2014, p. 88–90)
Knowledge	Factual information
Perceived Effectiveness	Judging a message to be convincing, compelling, and/or otherwise persuasive (Dillard, Shen, and Vail, 2007, pp. 467–468)
Recall	Remembering a message (either without assistant or with a prompt)
Story-Consistent Values	Identifying attitudinal, belief, or knowledge alignment between a story and audience members
Transportation	People feel like they are "entering the world of the story" (Hester and Schleifer, 2016, p. 109)

By connecting the cognitive power of stories coherently, narrative can engage individuals in the same way that real-life experiences do. This is why narratives are so powerful, because people automatically perceive the world with mental models that affect both what they perceive and how they interpret that incoming information (Davidson, 2017, p. 2). According to the Joint Staff J-7 focus paper on communication strategy and synchronization, a compelling narrative helps to promote the legitimacy of the mission and "prevent the 'say-do' gap in which our actions and words conflict in the eyes of the audience" (Joint Staff J-7, Deployable Training Division, 2016, p. 1).

These insights into how powerfully stories inform human cognition and interpretation are key to successful implementation of command narrative. In the next chapter, we explore practitioner-specific aspects of communication and narrative that we identified during site visits to the GCCs.

Figure 2.6
Assessment Measures and Indicators

MOE	MOP	Indicator
Answers the question, "Are we doing the right things?"	Answers the question, "Are we doing things right?"	Answers the question, "What is the status of this MOE or MOP?"
Measures purpose accomplishment	Measures task completion	Measures the data inputs to inform MOEs and MOPs
No hierarchical relationship to MOPs	No hierarchical relationship to MOEs	Subordinate to MOEs and MOPs
Often formally tracked in formal assessment plans	Often formally tracked in execution matrixes	Often formally tracked in formal assessment plans
Typically challenging to choose the correct ones	Typically simple to choose the correct ones	Typically as challenging to choose as the supported MOE or MOP

NOTES: MOE = measure of effectiveness; MOP = measure of performance.
SOURCE: Joint Staff J-7, Joint and Coalition Warfighting, 2011, p. III-5.

CHAPTER THREE

Practice in the Force for Command Narrative

Introduction

In the previous chapter, we laid out the findings and recommendations from a review of scholarly sources relevant to command narrative. In this chapter, we lay out our observations from subject matter expert interviews and site visits at the GCCs. First we explain our data selection frame and analysis, then describe narrative as understood by the force, and then devote the rest of the chapter to a detailed description of observations and best practices around command narrative and effective communications. We close the chapter with key findings from our analysis, and a set of recommendations aimed at the GCCs.

Data and Analysis

We conducted in-depth, semistructured interviews with 46 personnel across 31 organizations. The interviews were conducted in person or via telephone between September 2019 and April 2020. Interviews were transcribed but not audio-recorded. The majority of interviews were conducted with DoD civilians and military service members from the GCCs, as well as from the Marine Corps' Marine Expeditionary Force (MEF) Information Group (MIG) community.[1] We also spoke with individuals from the Office of the Secretary of Defense (OSD), the Joint Staff, functional combatant commands. and the military services. We also interviewed a range of military communication experts from PA, Information Operations, and Psychological Operations specialties.

Initial interviewees were selected based on suggestions by the project sponsor and by reaching out to individuals participating in relevant working groups or who had contributed to prior relevant reports and studies. Other potential respondents

---

[1] We note that the MIGs operate at the tactical and operational levels within theater commands, and thus are not directly comparable to GCC commands. We think insights from MIGs practice and organization are highly relevant to this study, however. The U.S. Marine Corps is at the vanguard of embracing information as a joint function and service warfighting function, and the rest of the joint force can benefit from its lessons learned.

were identified through referrals from interviewees (snowball sampling). To ensure an appropriate breadth of responses, we tried to obtain a breadth of operational and practical experience.

Our interview protocol was provided in advance to interviewees to give them an idea of what types of information we were interested in discussing. Interviews were conducted almost entirely by researcher pairs, with one researcher taking notes and the other acting as an interviewer. The protocol was used to structure the interview, with some questions emphasized more than others depending on the type of interviewee. The questions were grouped into several categories but centered on the definition of narrative, the creation of narrative, and the operationalization of narrative. Interviews were guided by the experience and knowledge of the interviewee, with the protocols used loosely to structure the conversation. This protocol appears in Appendix A.

After systematically collecting data through these semistructured SME interviews, the entire research team distilled and synthesized these data based on our own subject-matter expertise and experience, and our observations of the problem space. This included cross-team work to identify themes across interviews, and contrastive description of the different structures and processes we discovered across the joint force and at the GCCs. This chapter is the result of that analytic process and includes insights we concluded would be most useful for improving practice in the force.

Narrative

In the course of our interviews, we found a wide variety of practices around narrative, reflecting different underlying conceptions of narrative. Awareness of narrative ranged from rich, doctrinally informed concepts to confusion. Additionally, we found very different approaches to developing narrative across the GCCs and DoD supporting establishment.

Narrative Conceptions in the Joint Force

We can think of command narrative as a framework and overarching message strategy,[2] and the anchor for messaging activities.[3] It's the "party line," a way of grounding, "What are we saying we stand for?"[4] We found in our interviews within the GCCs and elsewhere in the joint force that this understanding is not widely shared.

---

[2]   Interview with experts in Public Affairs (PA) and Strategic Communications (StratComm), February 28, 2020.

[3]   Semistructured interview with expert in Communication Strategy and Operations (COMSTRAT), by phone, February 17, 2020; semistructured interview with expert in PA and StratComm, by phone, February 26, 2020.

[4]   Semistructured interview with Expert in Commander's Action Group, in person, February 28, 2020.

One alternative perspective was that "narrative" obscures the simple, straightforward idea of "stories" as the way humans understand the world (rather than *leveraging* stories in a military-specific way), with, for example, comparison to filmmaker Steven Spielberg, who tells stories that affect and engage the audience and *produce effects*.[5] Another perspective is that command narrative is simply the external version of the commander's intent, or even a redundant version of the commander's intent.[6] As one GCC staff section director put it, "commander's priorities are the narrative, [and we] don't need anything else."[7]

A somewhat different conception is that "[the commander's] vision statement is the command narrative. It should be one paragraph that paints a picture. It needs to be tied to the coordinated campaign plan."[8] In this conceptual model, narratives are geographically nested versions of the commander's vision statement in a pyramid: a master narrative on top, country narratives in the middle, and subregional and event-specific narratives at the bottom.[9] Another variation was the notion that command narrative is synonymous with cognitive dominance: "Command narrative means bending the will of the enemy. It is persuasion or coercion."[10]

Still other senior staff officers were simply unsure of what command narrative is, indicating that they neither understood it nor saw it as an important concept, arguing "Why do I need anything more than the commander's intent?"[11] Our literature review in the previous chapter showed a range of definitions for narrative in research that corresponds with the a lack of conceptual clarity we encountered in our interviews. This lack of a shared understanding reflects the joint force:

> A survey of how the concept is employed on popular blogs and web forums on defense issues suggests that "narrative" is a fuzzy notion as an instrument or practice. Generally, discussants are comfortable using the concept of narrative the way a historian might: to order and interpret past events, or to compare conflicting accounts. The "war on terror narrative," "Anbar narrative" and "surge narrative" have entered common usage. Narrative has become a synonym for "account" or "version of events." Members of the defense community are more vague when the discussion turns to narrative as an active function of information operations or public diplomacy. Narratives are described as amorphous concepts, such as

---

[5]   Interview with experts in COMSTRAT, via phone, February 17, 2020.

[6]   Semistructured interview with expert in Plans, in person, October 22, 2019; semistructured interview with expert in COMSTRAT, by phone, February 14, 2020

[7]   Semistructured interview with expert in Plans, in person, October 22, 2019.

[8]   Semistructured interview with expert in J-5/8 Comm Sync, in person, October 21, 2019.

[9]   Semistructured interview with expert J-5/8 Comm Sync, in person, October 21, 2019.

[10]   Semistructured interview with expert J-5/8, in person, October 23, 2019.

[11]   Semistructured interview with expert in J-35, in person, October 21, 2019.

"democracy" or "freedom" (e.g.: "We must learn how to tell our story of freedom better"). Alternatively, writers use "narrative" where one might otherwise use the term "propaganda," or "message." (Zalman, 2010, p. 4)

Narrative has also been confused with declarative, intentional language about what operations are meant to accomplish (Zalman, 2010, p. 4). And within one GCC, the term *command narrative* was used, but in a way synonymous with themes and specific messages.[12]

This confusion over narrative has powerful stakes for commanders as they seek operational success to serve strategic ends. Narratives makes sense of the world within our "ideology, theory, or belief, and . . . put things in their place according to our experience, and then tell us what to do. A strategic narrative aligns the strategy and the narrative so they become mutually supportive and integrated" (Laity, 2015, p. 23). Disparities in the joint force over perceptions about what narrative is and whether it matters point toward how important and challenging the task of operationalizing command narrative in the joint force will be. The next two sections address in detail the development and dissemination of command narrative.

An important element of the core of this chapter (and this report) concerns the development of narratives. Various practitioners with whom we spoke (or whose written observations we reviewed) shared their insights related to the development of narratives. Here, we present these insights in two sections: The first concerns the principles and practices recommended by practitioners, and the second details narrative development processes.

While this report is meant to broadly inform narrative across a range of commands, we acknowledge the need for a special focus on development at the theater level for the GCCs. GCC-level narrative development involves special complexity, in that it is more complicated because of the mix of audiences whose preexisting understandings and beliefs must be incorporated. It explains the command mission to the public and theater audiences, but needs to "incorporate an understanding of existing narratives, the frames that peoples use to view them, and how the command narrative nests under the U.S. or Coalition strategic narrative" (McGuinn, 2019, p. 21).

This is further compounded as commands navigate the same narrative terrain, operating distinctly and different levels, with distinct narratives (Multinational Information Operations Experiment, 2014, p. 11). NATO and EUCOM are two examples of such commands, with many specific differences in their command narrative operationalization, but with a shared landscape and ultimately a shared purpose.

---

[12] Group semistructured interview with Communication Synchronization Team, in person, December 23, 2010.

Practitioner-Provided Principles for Developing and Using Narrative

In describing the development of effective narratives, SMEs provided general principles they had derived or observed. These principles are either features that need to be in place prior to undertaking a specific process or postulates that guide a process throughout.

Commander Interest Is Key

We repeatedly heard that commander interest in narrative is essential to success. Simply put, if the commander doesn't care about narrative, it is hard to get anyone else in a command to particularly care, either. However, if narrative, or the information environment more broadly, is important to the commander and becomes a point of command emphasis, success ensues.[13]

Have a Forcing Function

One of the ways commander interest promotes effective narrative development is the role the commander plays as a forcing function. What the commander wants, the commander gets. A series of U.S. Southern Command (SOUTHCOM) commanders have emphasized narrative, and that emphasis has forced continued emphasis throughout the command.[14] During his tenure at SOUTHCOM, Admiral James G. Stavridis emphasized the importance of communication and narrative, saying variously, "Strategic communication is my main battery" and "I'm not here to launch missiles, I'm here to launch ideas."[15]

SMEs at EUCOM lamented a lack of forcing functions, wishing they had a stronger forcing function earlier in their planning cycle and noting that they are only forced to pay attention to narrative when they need one for an exercise or for theater campaign order revisions. By the time they are forced to have a narrative, that narrative has limited opportunity to shape anything about the exercise, plan, or order it is associated with.[16]

As noted, commander interest in narrative can be sufficient as a forcing function, but other forcing functions either in structure or process might also serve. Commanders who are less interested in narrative or who are interested but suffer from trying to provide attention across too many priorities might benefit from a forcing function themselves. Commander's intent can provide good guidance for and connection to narrative.[17] However, not all statements of commander's intent include or imply nar-

---

[13] Group semistructured interview with officer with experience with the Information Warfare Task Force in Afghanistan in 2019, by phone, March 20, 2020.

[14] Interview with experts in PA and StratComm, February 28, 2020.

[15] Quoted during interview with experts in PA and StratComm, February 28, 2020.

[16] Interview with experts in J-39, October 21, 2019.

[17] Interview with experts in PA and StratComm, February 28, 2020.

rative guidance. Some argue that all statements of commander's intent should do so. Dennis Murphy, while a professor at the U.S. Army War College, suggested that all statements of commander's intent should also include a commander's desired information end state, and one of the authors has echoed this suggestion repeatedly (Murphy, 2008).[18] A similar formulation leaves narrative separate from commander's intent, but always places narrative next to commander's intent at the core of operational design (McGuinn, 2019). Yet another way to force narrative into staff planning process is what NATO strategic communication expert Thomas Nissen has called "narrative-led operations," where both communications and kinetic actions undertaken by a command meet both strategic and narrative intent.[19] A final suggestion for a forcing function is Phil McGuinn's "keeper of the narrative" approach, whereby a specific staff section has responsibility for development of and advocacy for narrative throughout planning, execution, and in all staff processes (McGuinn, 2019). Any similar approach could provide a sufficient forcing function.

Nest with Higher Direction

Several SMEs reiterated the importance of narratives being consistent with direction from higher levels and key stakeholders. This includes guidance from OSD and the Joint Staff, existing command orders, strategies and plans, and the commander's intent.[20] This requirement for nesting applies to narrative and messaging at subordinate levels, where these lower-level efforts must nest with command or strategic narrative. One SME noted that, done right, the strategic narrative can shift or evolve without disrupting tactical or operational messaging.[21] This suggests that the principle of nesting imparts a responsibility both to those whose narratives should nest with higher guidance and to the higher guidance itself, to retain some narrative consistency over time.

A PA SME informed us that one effort to support nesting of themes and messages is the Joint Staff–provided Framework Messaging Guidance. Framework Messaging Guidance includes general guidance, themes to avoid, planning considerations, and information to help harmonize across different campaign plans.[22]

Narrative Needs to Be More Than Words, It Needs to Be Words and Actions

One insight provided to the Joint Staff J-7 by a combatant commander was that "narrative and messages are more than words—they include words and actions" (quoted in Joint Staff J-7, Deployable Training Division, 2016, p. 3). Numerous SMEs with

---

[18] Also see Paul, 2011.

[19] As described in McGuinn, 2019.

[20] Interview with EUCOM Future Ops, October 22, 2019; see, for example, Joint Staff J-7, Deployable Training Division, 2016, p. 3.

[21] Semistructured interview conducted with experts in Joint Operations, in person, February 27, 2020.

[22] Interview with PA expert, by phone, March 20, 2020.

whom we spoke echoed that point in one form or another, from "actions speak louder than words," to noting the importance of minimizing the say-do gap, to the importance of alignment between words and deeds. Actions and communications need to be consistent with intended narratives. However, this is not trivial; as Admiral Michael G. Mullen, former Chairman of the Joint Chiefs of Staff, asserted in 2009, "Leaders are not bad at communicating; they simply struggle with credibility and ensuring their actions match their words" (Johnson, 2011, p. 2).

Actions and communications should constantly be framed as contributing to the intended narrative and should "not be allowed to contradict it even though a particular tactical action or execution of a target of opportunity might be what wins the day" (Nissen, 2012). For example, one of the successes of information operations in the Iraq War between December 2007 and February 2009 was that "professionals gave guidance, and there was a close correlation between messages sent to the local population before any operation and the events that unfolded during and after any operation" (Schouten, 2016, pp 36–37). In contrast, the tactical actions of marines in Fallujah during Operation Vigilant Resolve were not properly synchronized with operations in the information environment, such as engaging a variety of Iraqi leaders, which contributed to the lack of support from the interim Iraqi government and the operation's failure (Metz et al., 2006, pp. 5–6).

Importance of Audience Perceptions

Another frequently mentioned guiding principle is the importance of audience perceptions of narrative. The command narrative is not primarily for internal audiences or oneself, so one must be mindful of who the narrative is for and how they will perceive it. A combatant commander highlighted the importance in thinking through how the narrative and messages will be perceived by the different audiences (Joint Staff J-7, Deployable Training Division, 2016, p. 3). Narrative must be designed to be seen through the lens of those for whom communicators hope it will have meaning.

Importance of Cultural and Linguistic Knowledge

It is not enough just to recognize that narrative is for one or more intended audiences; those audiences must be understood in order to address narratives to them and to estimate how those narratives will be received. "Creating these stories requires a deeper understanding of the relevant actors' cultural environment, the knowledge of the available stories and myths told in the local communities, and the networks within which they are told and by whom" (McGuinn, 2019, p. 23). Narratives and stories can be culturally specific, and developing a narrative for a specific audience requires a certain degree of cultural familiarity (Meier, 2016, pp. 9–10).

Recognize Different Levels of Narrative Opportunity

An article in the *Marine Corps Gazette* notes that in a given context and for a given audience there are often preexisting narratives, and the extent and strength of existing narratives will determine different levels of narrative opportunity (Paul, Colley, and Steckman, 2019, p. 83). Sometimes there is space to shape narrative or contribute new narrative, sometimes one can push in favor of one competing narrative over another, and sometimes there is only one dominant narrative for an event and you will not be able to move it (Paul, Colley, and Steckman, 2019). This notion of narrative opportunity echoes a point made in an earlier thesis from the Naval Postgraduate School: "Since narratives are neither fixed nor infinitely malleable, each side has a window of opportunity in which they may choose to change their narrative in order to address changing circumstances effectively" (Case and Mellen, 2009, p. 1).

Practices Endorsed by Practitioners

In addition to the various principles that practitioners helped us identify as a critical foundation to narrative development, various SMEs also contributed several positive practices that relate to or support these principles. We list and briefly describe these here.

Narrative practitioners note that the only way to beat a story is with a better story. Bad ideas and unfavorable narratives cannot be countered, but they can be "bettered" by a positive alternative narrative (McGuinn, 2019). Because it is hard to counter an accepted and engaging narrative, it is important to get the command's version of events out first. So, "Be first with the truth" (McGuinn, 2019). Doing so requires communication and narrative be proactive, and the narrative must be preplanned and released with timing coincident with actions or operations. This, in turn, requires that narrative be included early in the broader planning process.

Because actions speak louder than words and because there are often preexisting narratives likely to attach to certain actions, "Sometimes the only way to create an opportunity to change the narrative is to change the actions" (Paul, Colley, and Steckman, 2019, p. 82). Narrative is not just something used to justify a command's actions. If overall legitimacy is a goal and narrative helps to promote it, then sometimes actions and operations will need to be constrained in order to remain consistent with intended narrative. This is consistent with "narrative-led operations" and with commander's making narrative a priority (and part of commander's intent).

Practices to Be Avoided

SMEs also highlighted a number of practices that should be avoided in the design and dissemination of narrative. One SME noted that coercive narratives or messaging will not work in the long term and would require frequent repetition.[23] And, because the only thing that overcomes a story is a better story, communicators should avoid "parry" messages, which emphasize refutation or denial without offering a more compelling and engaging explanation for events (McGuinn, 2019).

---

[23]   Semistructured interview with expert in Information Operations, in person, February 27, 2020.

Narrative Development: Processes

This section focuses explicitly on processes for narrative development as identified through practitioner interviews and published notes on defense organization processes. Practitioners with whom we spoke highlighted and emphasized different aspects of process and described slightly different processes. Before detailing specific elements of process as described by various SMEs, we offer an overall summary. Synthesizing the various processes described suggests an overall process with these rough contours, depicted in Figure 3.1.

A narrative development process is likely to begin with some version of setting a preliminary objective or goal in terms of audience and objectives. Then, developers will need to understand the context, the environment, the audiences, and the things about them that might affect objectives or narrative (this understanding may be based on existing information and data, or it may require a new process or a subprocess). Next, based on this understanding, preliminary objectives will need to be reviewed to make sure they that are consistent with what is possible, to make sure that the right audiences have been identified, to add nuance to the objective, etc. With revised objectives and a good understanding of context, the next step is to develop preliminary candidate narratives, narrative elements, and other narrative material. This then leads to analysis, review, and evaluation of these candidate narratives to see how they fare against various design principles, guidelines, or other criteria. In later iterations, this evaluation step might include pilot testing, focus groups, or other formative evaluation methods.[24] This will likely lead to further iterative refinement of candidate narratives. What emerges from this iterative process will eventually be ready to move to dissemination (discussed later as a separate process). We note that developing a narrative is a time-bound process that must reflect operational realities: Iterations to refine narrative may need to be more like rapid prototyping and be "good enough," rather than perfect.

Figure 3.1
Generic Narrative Development Process

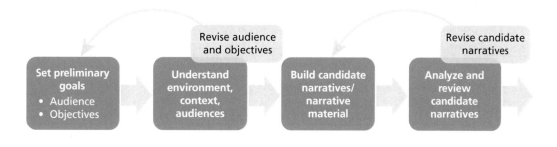

---

[24] For more on formative evaluation of defense efforts to inform, influence, and persuade, see Paul et al., 2015.

Elements of Process

The various elements of process identified through our SME interviews can all fit within the generic narrative design process depicted in Figure 3.1. Here, we list the various process elements that surfaced in our practitioner interviews under headings that correspond to the generic development process.

Set Preliminary Goals

Because narrative development is an iterative process that must start somewhere, it makes sense to begin with goals. Backward planning is a traditional approach to planning: Begin with the objective in mind, and then identify the path from the current location to that destination (Murphy, 2008, p. 4). These preliminary goals should include both the audiences to be addressed and the objectives relating to those audiences, which should ideally include some kind of attitudinal or behavioral change.[25] Though fairly self-evident, previous experience led one SME to prioritize starting with the desired objective before beginning to craft a message.[26]

Understand the Environment, Context, and Audiences

Preparing a narrative requires a fairly robust understanding of the relevant audiences (those people whose behavior is instrumental to the success or failure of the campaign); their narrative frames in terms of their history, worldview, and recent events; and the available narratives about the United States, U.S. forces, and their operations and actions. To be able to plan effectively for narrative, intelligence preparation of the information and operating environment must include attention to these kinds of issues. This preparation may require (or benefit from) media and social media monitoring or available behavioral, cultural, and linguistic SMEs who have sufficient knowledge of the operational context to meet the need (Paul, Colley, and Steckman, 2019). Understanding the language and context of an environment will help with recognition of what might make a resonant story and what might connect to existing unfavorable narratives (McGuinn, 2019, p. 23).

This element in the narrative development process seeks the necessary background and nuance to understand the context and anticipate how different narratives might be received. This will involve some combination of activities suggested by various practitioners, and it might look like the target audience analysis process found in doctrine for psychological operations or the audience segmentation and analysis process found in marketing. This part of the process might involve analysis related to the information environment drawn from intelligence preparation of the operational environment (IPOE), intelligence collection and analysis specific to narrative or to culture, or from social media monitoring. Contextual understanding should also review the communi-

---

[25] Semistructured interview with J-5/8 CSD, in person, October 21, 2019.

[26] Interview with experts in J-39, October 21, 2019.

cation environment to know what media and modes of communication are prevalent among relevant audiences in order to plan for dissemination.

As understanding of the context and audiences grows, that should lead back to the first process element, the specification of objectives. Understanding may lead to revised objectives, either in terms of desired behaviors or in terms of specific audiences or segments to be addressed.

Build Candidate Narratives

The practitioners with whom we spoke had little to say about the practice of actually drafting narratives. Clearly that has to happen at some point, and the implication in our interviews is that it is as much art as it is science. Regarding the science, see Chapter Two. At the intersection of art and science is the fact that narrative is at least partially rhetoric and designed to persuade someone of something (Simpson, 2012, p. 188). Within that framework, several sources suggest the Aristotelian approach: Develop narratives with appeal to logos (rational argument), pathos (emotional appeal), and ethos (moral standing).[27] Of the three, pathos may be most important. Kendall Haven's pathbreaking research on effectiveness in persuasive storytelling notes the emotional resonance of a story as the single most important factor (Haven, 2014).

NATO's Mark Laity offers what he calls the "narrative arc" approach to laying out a narrative. Under this approach, one lays out four steps (the arc) in sequence and builds a narrative that encompasses the whole arc. The four components he notes are (1) the problem or conflict, which leads to (2) the desire to resolve or solve it, suggesting (3) a series of actions to be taken, ultimately resulting in (4) a desired resolution (Laity, 2015, p. 27).

The NATO Strategic Communications Centre of Excellence has instructional materials on developing narratives and offers that a narrative should include four elements: the current state, a future state, a pathway, and a justification.[28] When developing narratives, the center recommends, one should try to identify candidates for each of these elements that are consistent with what one hopes to accomplish.[29]

Review and Refine Proposed Narratives

Once contextually informed draft narratives are completed, the next process element is to evaluate and refine those narratives. Some of this evaluation should take the form of comparing proposed narratives to various lists of criteria. This could be a check to see whether the proposed narrative has each of Aristotle's logos, pathos, and ethos; Laity's complete "narrative arc" (a problem, a desire to resolve it, actions to be taken, a desired resolution), or the NATO Strategic Communications Centre of Excellence's four ele-

---

[27]   See Simpson, 2012, p. 188; Casebeer and Russell, 2005, p. 8.

[28]   Interview with personnel from NATO Strategic Communications Centre of Excellence, March 9, 2020.

[29]   Interview with personnel from NATO Strategic Communications Centre of Excellence, March 9, 2020.

ments (current state, future state, pathway, justification). Similar evaluative criteria might stem from the material in Chapter Two, or from the detailed processes discussed in the next section.

Initial reviews and evaluations are likely to find that preliminary candidate narratives are deficient in some areas, which pushes the process back into another iteration of building and revising candidates. In later iterations, this evaluation step might include pilot testing, focus groups, or other formative evaluation methods.[30]

Candidate Processes

In addition to elements of process drawn from SME interviews and practitioner documents, some sources propose whole processes for narrative development. This section briefly summarizes these. Note that most fit overall into something like the generic three-step iterative process described above, but with offer greater granularity in substeps or sub-elements.

Seese and Haven Story-Building Process

Gregory Seese and Kendal Haven (2015, p. 35) propose a process in what they call the Storyline Narrative Methodology:

1. target audience analysis
2. influence message (theme statement and objectives), with five substeps:
    a. desired outcome/behavior
    b. knowledges and beliefs
    c. theme message
    d. residual resolution emotion (RRE)
    e. check/confirm
6. metaphoric image
7. context and relevance check
8. real-world constraints
9. identify key character positions
10. build the story.

Haven's Seven-Step Process

Prior to his collaboration with Seese, Haven proposed a slightly less detailed and refined version of the same process (Haven, 2014, p. 140):

1. Define the target audience.
2. Create your theme and "take-away" message.
3. Search for a core metaphor and image.
4. Create relevance and context.

---

[30] For more on formative evaluation of defense efforts to inform, influence, and persuade, see Paul et al., 2015.

5. Adjust for real-world constraints.
6. Define/develop story characters.
7. Build the story elements.

Multinational Information Operations Experiment Process Elements

Draft guidance on developing narratives for coalition operations from the Multinational Information Operations Experiment offers some additional considerations related to the narrative development process (Multinational Information Operations Experiment, 2014). It urges consideration of

- Purpose: Why to be done?
- Input: What information is to be used and/or considered?
- Procedure: What activities are required?
- Involvement: Who needs to contribute?
- Output: What products shall be delivered?
- Tools and templates: What instruments shall be used?

The Army Asymmetric Warfare Group's Process

The Army Asymmetric Warfare Group (AWG) advocates a process that fits within the mission analysis step within the broader joint operations planning process or military decisionmaking process. The AWG method begins with seeking to understand the operational environment, then moves to building the narrative, then delivering the narrative, then assessing effects. This process is depicted in Figure 3.2.

UK Ministry of Defence OASIS Model

The UK Ministry of Defence promotes application of the OASIS model from industry to defense contexts. OASIS stands for objective, audience insights, strategy formulation, implementation, and scoring. The way this model applies in the defense context is elaborated in Figure 3.3.

Narrative Dissemination

Once a narrative has been developed, it must be disseminated; a narrative does no good if it just stays within a staff section of a GCC. The questions quickly become: Who needs to know the narrative, and how does that narrative get to its intended audiences? To answer these questions, it is essential to understand the key audiences. For a GCC, there are both internal and external audiences, which can vary depending on how broadly one scopes the audiences. In general, though, the focus for a command narrative is on external audiences.[31] Figure 3.4 shows an example of internal and external audiences at the GCC level.

---

[31] Semistructured interview with Russia Strategic Initiative, in person, October 23, 2019.

Figure 3.2
The Army Asymmetric Warfare Group's Narrative Methodology

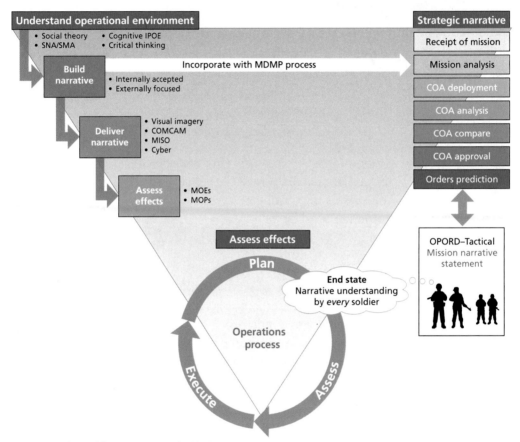

SOURCE: Adapted from Moore et al., 2016.
NOTES: IPOE = intelligence preparation of the environment; SNA/SMA = social network analysis/strategic
multilayer assessment; MDMP = military decision-making process; COA = course of action; COMCAM =
Combat Camera; MISO = military information support operations; MOE = measure of effectiveness;
MOP = measure of performance; OPORD = operation order.

Internal audiences include the commander, all staff elements, subordinate commanders and their staff, the Joint Staff, and OSD. External audiences include other governmental agencies and organizations, including the executive and legislative branches; partners and allies; adversaries; neutral populations; and the American public. However, not all of these audiences receive or "hear" the narrative in the same fashion. Stories are more likely to be received and understood if they are fit into a person's worldview and in a manner that is acceptable and understandable. To be effective, a narrative's stories should take advantage of heuristics and biases (for example, memorable events will be easier for a target audience to recall and could therefore be the basis of a story) (Casebeer and Russell, 2005, p. 8). Therefore, not only is the content important, but so is the lens through which a person receives the message. This

Figure 3.3
The OASIS Model as Advocated by the UK Ministry of Defence

OASIS heading	Defence equivalent	Plain English equivalent
Objective	Define military strategic objectives.	Construct Military Strategic Effects around the behaviours to be sought from target audiences.
Audience insights	Target audience analysis.	Gain sufficient understanding of the target audiences and how they communicate to be able to influence them effectively.
	Information environment analysis.	
Strategy formulation	Define strategic intent; construct Defence strategic narrative.	Formulate Defence strategy around the activities most likely to be effective in delivering the required behaviours. Deduce effects and activities. Issue direction.
	Identify effects; extrapolate activities; construct Chief of Defence Staff's Directive.	
Implementation (execution)	Execute strategy to delivery activities and generate effects.	Execute strategy and monitor whether activities are generating the intended outputs.
	Manage activities as the situation evolves.	
Scoring (evaluation)	Measure effectiveness of activities.	Monitor target audiences for evidence of desired behaviours. Adjust activities in concert with other government departments.
	Adjust activities.	

SOURCE: UK Ministry of Defence, 2019.

takes form through a variety of methods. It can be written, spoken, conveyed through action or inaction, attributed or intentionally not attributed. Depending on the audience, the message can change form, tone, or other characteristics.

For internal audiences, the narrative must be an executable document. Ideally, it should be published in either the Theater Campaign Order or a Consolidated Campaign Plan. As one interviewee noted, "A narrative is just a piece of paper unless it's executed."[32] Once published, the narrative can (and should) be reinforced through a variety of mechanisms for internal consumption. Meetings of groups such as the CCSWG (Commander's Communication Synchronization Working Group), Information Operations Working Group (IOWG), Target Development Working Group (TDWG), and Joint Targeting Coordination (or effects) Board (JTCB) should all at least mention the narrative on a regular basis. Newsletters, point papers, and talking points are also mechanisms to "push" the narrative.[33] Communications professionals

---

[32] Semistructured interview with Training Division, in person, October 24, 2019.

[33] Semistructured interview with Russia Strategic Initiative, in person, October 23, 2019.

Figure 3.4
Internal and External Audiences Example

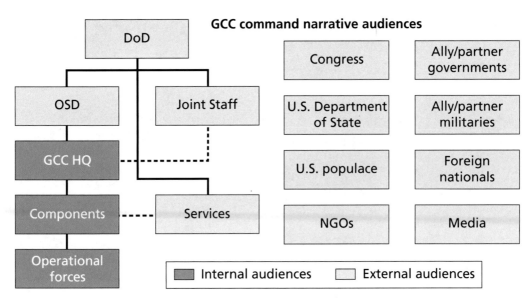

NOTE: This figure provides examples of what may be some internal and external audiences (it is not intended to be all-inclusive, since every situation varies). NGOs = nongovernmental organizations.

across the staff are also well suited to highlight the narrative through a variety of their routine staff outputs, such as the generation of public affairs guidance, communication plans, press releases, social media posts, responding to queries, working with Congress, speeches, and sharing transcripts after the fact.[34]

For external audiences, how command narrative is executed is more difficult, because it cannot be controlled like it is for internal audiences. As one interviewee noted, after it is planned, narrative is essentially handed off for another entity to execute.[35] For example, one GCC uses a variety of local and international media outlets to get its narrative out. According to one press coverage roundup, the command leadership was quoted in domestic and international newspapers, online and throughout multiple countries within the GCCs area of responsibility.[36] Commands can tell a story, but if the story is not compelling and interesting to external audiences, it will

---

[34] Semistructured interview with expert in PA, in person, October 21, 2019.

[35] Semistructured interview with Future Ops, in person, October 22, 2019.

[36] SOUTHCOM media roundup, provided to research team by SOUTHCOM communications personnel, January 30, 2020.

receive relatively little traction. Simple tactics, such as including imagery, can help a story to trend on social media, but this alone will not help push a narrative.[37]

Finally, it is a mistake to think that narrative is solely or primarily about messaging. Messages and deeds must be aligned to minimize the say-do gap. Remember that sometimes planned communication needs to change to conform to planned actions, but sometimes planned actions need to change to conform to communication goals. "Sometimes the only way to create an opportunity to change the narrative is to change the actions" (Paul, Colley, and Steckman, 2019, p. 82).

Communication Coordination

One of the key research aims of our study was to identify best practices in communication coordination—ensuring that messaging is coordinated and harmonized, and also in accord with actions so there is no say-do gap. JP 1, *Doctrine for the Armed Forces of the United States*, requires such coordination, noting that "Staffs develop the approach for achieving information-related objectives and ensur[e] the integrity and consistency of themes, messages, images, and actions to the lowest level through the integration and synchronization of relevant information-related capabilities" (JP 1, 2017, p. I-13).

Close Collaboration and Synchronization Versus Deconfliction

We identified two models for communication coordination in our site visits and interviews: a *close collaboration and synchronization* approach and a *deconfliction* approach.[38] Close collaboration and synchronization was marked by commander engagement, structural and process integration across staff functions (including the Public Affairs Office [PAO] in all aspects of communication), and a shared understanding of the command's narrative. Deconfliction, on the other hand, was marked by commander disengagement from communication beyond PA, distinct communication efforts with post-hoc deconfliction of completed plans, a firewall between PAO and many other communication efforts, and no shared (or any) understanding of command narrative.

Practitioners referred to their local practices by a local idiomatic term (*comms integration*) but did not refer to the joint doctrinal process of commander's communication synchronization (CCS). When examined, the preferred process we refer to as "close coordination and synchronization" was more like the CCS process. CCS requires the

---

[37] Group semistructured interview with officer with experience with the Information Warfare Task Force in Afghanistan in 2019, by phone, March 20, 2020.

[38] Interview participants actually used different language to characterize these two approaches (for example, using *integration* to described what we call *close collaboration and synchronization*). We have changed the terminology to both preserve the essence of the approaches and use language in ways consistent with usage in our primary audience (*integration*, for example, usually refers to a structural arrangement in DoD parlance, whereas *close collaboration and synchronization* could be the product of either structural integration or effective processes).

commander's involvement early in the process, is integrated in the full staff planning process, and aligns the GCC's actions from the U.S. national level to the tactical level (JP 5-0, 2017, p. III-10). The process implied in what we refer to as "deconfliction" could be viewed as an absence of the CCS process on the GCC staff. We detail these two distinct models for communications in the following sections, after first reiterating the importance of commander engagement.

Commander Engagement

The starkest contrast we saw between the close collaboration and synchronization model and the deconfliction model for communication coordination was commander engagement. In the GCC we visited that used a close collaboration and synchronization communication model, the commander was directly involved in communications planning and made communicating the command's story—the command narrative—a priority. This direct command involvement raised attention to communication for the entire command. As one PAO chief put it, "When strategic communications is a priority of the commander it quickly becomes a priority for the entire command and those charged with directing strategic communication efforts are better equipped to ask for resources and personnel when they have the support of the CCDR" (Aguilar, 2011, pp. 16–17).

Structure and Process for the Close Collaboration and Synchronization Model

Again, within the GCCs we found two distinct sets of structures and processes for communication coordination: a close collaboration and synchronization model and a deconfliction model. The former was primarily focused on an integration mindset and processes to ensure synchronization of communication, whereas the latter sought only to deconflict largely independent efforts. A parallel structural issue we uncovered was over the level of involvement of PA in wider communications practice. We detail these distinct models below.

At one GCC, the close collaboration and synchronization model was explicitly labeled *comms integration*, and staff at this GCC made a point of explicitly saying they did "integration, not synchronization."[39] By this, they meant that communication coordination was baked in structurally and in all processes, not applied post-hoc. Synchronization in this model started with commander, who attended and ran a bi-monthly scheduling and communications meeting. Also at this GCC, a one-star general officer leads daily coordination calls that are interagency and span the area of operations and senior leader levels.[40] These meetings poll the entire staff and include discussion of long-range planning, planned media message, planned testimony, and

---

[39] Interview with experts in PA and Strategic Communications (StratComm), February 28, 2020; semistructured interview with expert in Commander's Action Group, in person, February 28, 2020; group semistructured interview with experts in Joint Planning, in person, February 28, 2020.

[40] Semistructured interview with expert in Joint Operations, in person, February 28, 2020.

audience expectations. There is an additional Strategic Communications (StratComm) Working Group,[41] co-led by both the StratComm chief and PAO, that meets weekly, with a more tactical focus on specific themes, events, and key leader engagement (KLE) activities.

Another example of baked-in communication synchronization at this GCC was the ubiquity of command narrative and themes derived from the command's narrative. We were struck by how, across staff functions, we heard the same framing language: unprompted, members of PA, StratComm, the Commander's Action Group, the Foreign Policy Advisor (POLAD), and J-2, J-3, and J-8 all brought up and used the common command narrative as a way to contextualize and frame what they were doing. This ubiquity was vertical as well as horizontal: Staff officers and senior civilians were all familiar with the command's narrative, and we found that staff noncommissioned officers and even junior noncommissioned officers were familiar with the command's narrative.[42] Additionally, multiple personnel at this GCC framed operations as being a kind of message nested within the command narrative, even pointing to operational names as reflecting the command's narrative. In essence, *command narrative functioned in this GCC as it is meant to: as a way to ensure coordinated communication and action to meet theater campaign objectives.*

GCC staff emphasized to us that the reason that this model works is because "it has GOFO [general officer/flag officer] power behind it"—that is, because the commander prioritized coordination and was personally involved, the staff followed the commander's lead.[43] This theme of commander's priorities and communication synchronization as "personality-driven" was echoed across staff sections.[44] Meaningful collaboration for communications may require a commander's full buy-in.

We note this this model has traction beyond the joint force. In our interviews at the NATO Strategic Communications Centre of Excellence, experts there said the aspiration was to have a StratComm representative in all branches of the staff process, such that each headquarters staff has a StratComm representative on all its boards, bureaus, centers, cells, and working groups (B2C2WGs).[45] They also explained that their goal was to inculcate a mindset across the staff that recognizes that everything the force does creates an imprint in the information environment, and also to coordinate impact on that environment through synchronization of communication efforts.[46]

---

[41]   Although *strategic communications* is no longer a U.S. doctrinal term, this GCC still uses it.

[42]   Group semi-structured interview with experts in Joint Intelligence, in person, February 28, 2020.

[43]   Interview with experts in PA and StratComm, February 28, 2020.

[44]   Group semi-structured interview with experts in Joint Intelligence, in person, February 28, 2020; group semi-structured interview with experts in Joint Planning, in person, February 28, 2020; semistructured interview with expert in Joint Operations, in person, February 28, 2020.

[45]   Interview with personnel from NATO Strategic Communications Centre of Excellence, March 9, 2020.

[46]   Interview with personnel from NATO Strategic Communications Centre of Excellence, March 9, 2020.

The Deconfliction Model: Post-Hoc Coordination Between Stovepipes

Other GCCs have a communication deconfliction model: Communication development is more siloed, with structures and processes meant to deconflict those disparate processes. For example, one GCC had an assortment of meetings and working groups, some active and others notional, and what coordination and synchronization there was spread out across different events and between different parts of the staff.[47]

In this example, PAO had its own separate dialing meeting to develop a PA playbook with attendant themes and talking points.[48] The Information Operations Working Group (IOWG), meanwhile, had a weekly meeting, but without PA representation, while the Key Leader Engagement Working Group had not met for almost a year and had abandoned the use of their KLE tracker.[49]

GCCs using a communications deconfliction model did not have command narrative statements. Instead, primary coordination responsibilities lay with a CCSWG. At one GCC, the CCSWG had fairly broad attendance across staff functions, including representatives from J-0, J-2, J-3, J-5, and J-7; however, PAO staff did not attend. The primary activity though was not synchronization, but rather theme development (independent of PAO) to be deconflicted later. This CCSWG was described as a "coalition of the willing" without a forcing function, and one SME noted that while staff functions had two-star leads, the CWSG was chaired by a one-star official.[50] A variation at another GCC was a communications synchronization team (CST), whereby the same representatives attend to try and build up personal continuity with weekly strategic alignment discussion (SAD) meetings, with similar makeup to CCWSG.[51] The CST was responsible for keeping various teams across the command on the same page (deconfliction) and facilitating campaigns. This structure was described as "semipermanent," and CST and those that participated in SAD meetings struggled with the fact that there was a lack of binding guidance.

What we observed in these GCCs reflects broader understanding in the joint force around communications best practices (Joint Staff J-7, Deployable Training Division, 2016, p. 9). One idea is that communication-related working groups closely coordinated with operations can achieve effective synchronization, to ensure that actions match words and avoid what some call "effects fratricide." Another idea is that designation of an Office of Primary Responsibility for communication and the assignment of staff responsibility for planning, coordinating, preparing, monitoring, post-engagement debriefs, database structure, recording, dissemination, and follow-up

---

[47]   Group semistructured interview with experts in Joint Planning, in person, February 27, 2020.

[48]   Group semistructured interview with experts in PA, in person, February 27, 2020.

[49]   Group semistructured interview with experts in Joint Planning, in person, February 27, 2020.

[50]   Group semistructured interview with experts in Joint Planning, in person, February 27, 2020.

[51]   Group semistructured interview with communication synchronization team, in person, December 23, 2010.

actions. Neither of these ideas is anchored in command narrative, allowing for close collaboration and synchronization, and thus do not appear to be actual best practices.

Synchronization Versus Deconfliction for Geographic Combatant Commands

One of our main findings, then, is that there is wide variability in how combatant commands organize for and understand the broader issue of coherency in communications. One approach to how the GCC communicates across a range of activities is closely collaborative and synchronized, and anchors on the command's narrative. A different approach is to create structures and processes that seek to coordinate and deconflict essentially distinct communication efforts, without attention to command narrative. The latter approach was fraught with difficulty: lack of engagement, inconsistent participation, failure to collect data systematically, and disagreement about communication ends and means between staff. We note that combatant commands have great flexibility with staff structure and process, and that commanders have the freedom to choose what they emphasize and how they conduct operations. We highlight that this choice has serious implications for coherence of command's communication efforts, and that a communication synchronization model better supports that coherence.

Public Affairs Office Involvement: Two Distinct Coordination Models

The contrast in PAO involvement between these two communications models is a particularly bright distinction. We found that full PAO involvement was foundational to broader communications synchronization, whereas in the deconfliction model, PAO hid behind a firewall meant to insulate PAO from other information-related activities.

Public Affairs Office Participation

One approach to PAO participation is full involvement, such that PAO is intimately involved across the staff and is a close collaborator in communication efforts.[52] PAO collaboration is important because it increases competition capability, helps create unity of effort, and complements long-range strategic communications with shorter-term tactical messaging.

In terms of communications capability, PA has a unique, powerful affordance in terms of range: PA can talk to multiple audiences (domestic, international, and adversarial) with requisite authorities and permissions. PAO's unique reach and credibility is particularly critical for communicating effectively with international audiences (Aguilar, 2011, pp. 1–2). As one PA expert put it, communication efforts across staff functions need PAO's capability to engage multiple audiences: "In the competition state,

---

[52] Interview with experts in COMSTRAT, via phone, February 17, 2020; interview with experts in PA and StratComm, February 28, 2020.

PA is 75 percent of your capability. If your PA folks aren't in the room, you will never win."[53]

PAO can also offer an important role in terms of vetting and coordinating other informational activities and breaking down these firewalls enables commanders "to produce a more unified SC [strategic communications] effort" (Aguilar, 2011, p. 7). According to Brian Cullin, former Director for Communication Planning and Integration at OSD for PA, "Not integrating [PA] is like not including an awareness of the battle space when planning an operation" (Aguilar, 2011, p. 15). This unity of effort can be further enhanced through complementary tempo effects. PA has a day-to-day tactical focus, whereas joint planners and strategic communicators have a longer-range, strategic focus, and thus collaboration and synchronization supports both daily and longer-term communication efforts across the command.[54]

Another hallmark of this approach is close working relationships between PAO and the J-2. Staff members emphasized the personal relationship between PA and J-2 staff, whereby the PAO chief routinely came to the J-2 to coordinate with open source branch members and J-2 targeteers.[55] In turn, the J-2 targets chief is a standing member of the command's Strategic Communications Working Group, feeding intelligence assessments back into the communications process.[56]

Public Affairs Office Firewall

In contrast, GCCs with a communication deconfliction approach were marked by a clear firewall: PAO functioned independently of other information activities, was defined by special staff function, and explicitly avoided specific operational knowledge so as to have plausible deniability.[57] This is the dominant model in the force, and this organizational gap between various functions is visualized in Figure 3.5. These organizational gaps may lead to coordination and planning gaps (Baldwin, 2007, p. 15), threatening to increase say-do gaps.

PA experts within GCCs that emphasize the PAO firewall also emphasized their special relationship with the commander and direct reporting nature, pointing to the commander's reliance on PA to take a variety of inputs (National Defense Strategy, commander's priorities for messaging, and OSD, the Department of State, and the

---

[53] Interview with PA expert, via phone, March 20, 2020.

[54] Semistructured interview with J-5/8, in person, October 23, 2019; group semistructured interview with experts in Joint Planning, in person, February 28, 2020; interview with experts in PA and StratComm, February 28, 2020.

[55] Group semistructured interview with Experts in Joint Intelligence, in person, February 28, 2020.

[56] Group semistructured interview with Experts in Joint Intelligence, in person, February 28, 2020.

[57] Interview with experts in COMSTRAT, via phone, February 17, 2020; interview with experts in COMSTRAT, by phone, February 17, 2020; interview with experts in PA and StratComm, February 28, 2020; interview with personnel from the NATO Strategic Communications Centre of Excellence, March 9, 2020; semistructured interview with expert in PA, in person, October 21, 2020.

Figure 3.5
Firewall Between Inform and Influence Staff

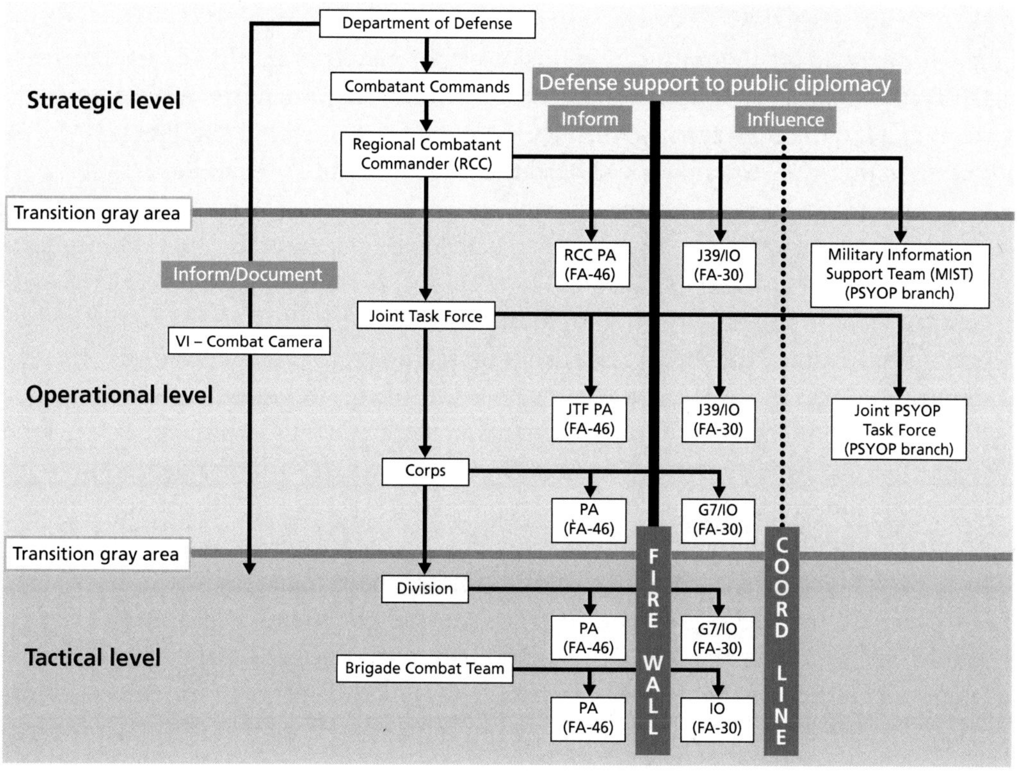

SOURCE: Baldwin, 2007, p. 15.

Joint Chiefs of Staff) and agilely produce a slew of outputs at a high tempo.[58] Such products include

- themes and messages, component command posture and requirements, and expected questions and guidance.
- regular communication with components to coordinate and reinforcing messaging
- commander's talking points
- responses to external queries
- posture updates
- daily news clips to staff
- ad hoc sentiment reports on various issues,
- communication plans

---

[58] Group semistructured interview with experts in PA, in person, February 27, 2020; semistructured interview with expert in PA, in person, October 21, 2020.

- social media presence
- "key top lines" from general officer/flag officer public speaking engagements.

Tearing Down the PAO Firewall

Synchronization of communications and collaboration of communicators (particularly PAO) into command-wide communication efforts such as command narrative offers many potential advantages. Within the GCCs, but also within and beyond the joint force, organizations are moving toward the integration of PA functions.

The Marine Corps has torn down the PAO firewall: PA has been merged with Combat Camera to create a new military occupational specialty for communication strategy and operations (COMSTRAT). Each of the three MEFs has a MIG with a company-sized element, creating a tactical unit with PA capabilities, integrated with other informational capabilities. Instead of remaining at arm's-length from other communication efforts, COMSTRAT marines are integrated into the MIG's Information Command Center (ICC), a command center analogous to a Combat Operations Center (COC), integrating cyberspace, electronic warfare, military deception psychological operations, COMSTRAT, and intelligence.[59] As one Marine Corps SME put it, integration of PA functionality into the ICC "provides greater situational awareness and puts [us] in a better position to communicate a command narrative."[60]

The erosion of PAO firewall extends beyond the joint force to NATO as well. While NATO allows that the two functions need to operate separately and have different responsibilities and approaches, NATO seeks to integrate all communication efforts in one branch.[61] Keeping all communications efforts together is intended to enable full deconfliction of efforts and synchronized effects. Additionally, one GCC is in transition, moving toward a more integrated and synchronized model for command communications; however we note that the PAO at this GCC expressed opposition to this, characterizing nascent attempts at collaboration as "recreational communications" and complaining that PA efforts were being weighed down by the attempt to synchronize across communication efforts.[62]

We point to other examples of increasing PAO involvement across the joint force. For example, 16th Air Force has created an Information Warfare Cell in which PA planners work intimately to support the command's operations in support of GCCs. Additionally, prior to its cancellation due to COVID-19, the 2020 exercise DEFENDER '20 planned to employ a Theater Information Command with a two-star commander

---

[59]   Interview with experts in COMSTRAT, by phone, February 14, 2020.

[60]   Interview with experts in COMSTRAT, by phone, February 14, 2020.

[61]   Interview with personnel from the NATO Strategic Communications Centre of Excellence, March 9, 2020.

[62]   Semistructured interview with expert in PA, in person, December 23, 2019.

that would have incorporated PA as it provided an information component to support EUCOM and the Joint Task Force Commander.[63]

Given the problems endemic to a PAO firewall, and the example of NATO, the Marine Corps, and some of the GCCs, it likely makes sense for commanders to consider tearing it down as they move toward closer communication collaboration and synchronization. This is not to say that PA is not meaningfully distinct from other communication efforts in some ways, and there may be arguments in favor of maintaining some aspects of the PAO firewall. It is important that PA messages remain completely truthful and that PA credibility is protected. PA experts in commands that emphasize a PAO firewall reported to us in interviews that they are the real professionals in communication, that they have a particular obligation as truth-tellers (and that other capabilities threaten that truth-telling function), and that they highly value their special relationship with the commander.[64] We think it is possible to respect many of these concerns while still reaping the benefits of closer communication collaboration. Having situational awareness of other communication and information efforts in the command does not compromise PAO—on the contrary, improved situational awareness will help PAO avoid unintentionally delivering misinformation.

Assessment

Assessing narrative and its effects is both extremely important and extremely difficult.[65] Perhaps the most difficult part of assessment is linking narrative to campaign objectives,[66] but assessment succeeds or fails on the efforts assessed having clear and measurable objectives—otherwise, it is not assessment (Paul and Matthews, 2018). Narrative assessment can be done, and there are good examples in other domains from which to draw. For example, debate surrounding the Affordable Care Act in 2017 involved dueling narratives. McGuinn gives an example of assessing narrative: "The Republican narrative focused on 'sky-rocketing costs' while the leading Democratic narrative 'playing politics with patients' and generated, respectively, 2% and 56% of the conversations on-line" (McGuinn, 2019, p. 14). While online measures would be only part of a robust assessment, it is an example of a clear measure linked to a campaign objective of audience adoption and spread.

---

[63] Email correspondence with expert in COMSTRAT, October 20, 2020.

[64] Group semistructured interview with experts in PA, in person, February 27, 2020; semistructured interview with expert in PA, in person, October 21, 2020.

[65] Group semistructured interview with experts in Joint Assessment, in person, February 28, 2020; group semistructured interview with experts in Joint Intelligence, in person, February 28, 2020.

[66] Group semistructured interview with experts in Joint Assessment, in person, February 28, 2020.

An Example of Narrative Assessment from the Force

Generally, in our site visits to the GCCs, communication assessment was nascent or missing. We did observe data collection that informed some messaging, but not collection and analysis to assess the effects of messaging. For example, at one GCC there was limited social media monitoring—for example, to inform PA sentiment reports and open source analysis by the J-2—but data collection for effects of messaging was generally absent or disconnected from planning.[67] This is not to say that staff members at GCCs we visited did not see the value of assessment related to narrative, just that activity in this area was limited or nascent.[68] This lack of assessment may reflect the inherent difficulty in assessment. As a process, cognitive effects are hard to understand and see, and staff at that GCC expressed that staff disagreement over what and how to assess effects was a serious impediment.[69] At another GCC, staff pointed out that communication assessment was absent because it was too time-consuming and required additional manpower.[70]

A Deliberate Assessment Effort Anchored in Communication Synchronization

This difficulty of assessment highlights how important it is to include assessment as part of a synchronized communication plan oriented around command narrative. The one GCC we observed that employs a communication collaboration/synchronization model is explicitly building assessment capability, marked by structured data collection and the incorporation of business analytics. We note that assessment of narrative and campaign plan effects in this case was a work in progress. Building a robust assessment capability is a challenging task, because assessment requires clear and specific objectives at both the campaign and command narrative levels, which in turn requires precise knowledge about audience and specific measurables for influence through behaviors.[71] Experts within the J-8 at this command said that the goal of their assessment effort was to gather and integrate structured data relevant to campaign objectives, conduct analytics to find "executive level information to inform corporate decisionmaking," and thus present the commander with a "single version of the truth."[72] Though these assessments are produced by personnel within J-8, they draw on monitoring and data collected by J-2, PA, the Information Operations staff, and possibly J-5 (depending on the exact data required).

---

[67] Semistructured group interview with J-39 Assessment, in person, October 22, 2019; interview with experts in J-39, October 21, 2019; semistructured interview with expert in PA, in person, October 21, 2019.

[68] Semistructured interview with Training division, in person, October 24, 2019.

[69] Semistructured interview with J-5/8, in person, October 23, 2019.

[70] Group semistructured interview with Communication Synchronization Team, in person, December 23, 2010.

[71] Group semistructured interview with experts in Joint Assessment, in person, February 28, 2020.

[72] Group semistructured interview with experts in Joint Assessment, in person, February 28, 2020.

Using Data and Analytics to Assess Narrative Effects

This particular assessment process started with an effects-based plan to link operations to theater objectives. Specifically, this includes 42 objectives and more than 600 individual measurements. To handle an effort that large and complex, the command uses business analytics to streamline the collection and processing of big data to answer its assessment questions.[73] A specific dimension of this command's assessment process is the use of structured data aggregated in permanent databases. For example, the commander's calendar is hosted on a classified network using web-based software that includes drop-down menus and semistructured data fields to capture relevant KLE visit information, and this information automatically populates a database.[74] The J-5 builds the commander's book for visits using this system, which includes talking points, intelligence estimates, and other country-specific information for the commander.[75] Using this web-based software, everything from KLE visit after-action reports, to themes and messages, to exercises are entered into a query-able database. This in turn allows for a large-scale, annual assessment of the command's campaign, specially anchored in the campaign-level structure of its command narrative.

Both J-8 uniformed and civilian staff emphasized the criticality of data.[76] They argued that assessment, as a form of analytics, had to be grounded in data. Further, they stressed that long-term trends and insights can only be visible if data are collected and maintained over time, so that collection efforts and datasets survived the rotation of staff. Finally, they also stressed the need for coordination and cooperation between staff sections to support data analytics.

How Intelligence Supports the Communication Integration Process

The intelligence function has a large role to play in developing narrative. The U.S. Intelligence Community has the responsibility to understand threat actors, actions, and motives, but is also responsible for understanding sociocultural factors affecting the operating environment (JP 2-01.3, 2014, p. VII-2). Part of these factors are the social structures and cultural factors, such as attitudes, perceptions, and beliefs. These elements are critical to understanding an audience, which is a precursor to crafting an effective narrative that resonates with them. However, with limited resources and competing priorities, the intelligence community may not be able to produce this kind of analysis across large populations and audiences (Schwille et al., 2020). This is not to say that the Intelligence Community cannot provide assistance in narrative develop-

---

[73] Group semistructured interview with experts in Joint Assessment, in person, February 28, 2020.

[74] Semistructured interview with expert in Commander's Action Group, in person, February 28, 2020.

[75] Semistructured interview with Commander's Action Group, in person, February 28, 2020.

[76] Group semistructured interview with experts in Joint Assessment, in person, February 28, 2020.

ment, for indeed it can. Specifically, when properly leveraged, the Intelligence Community can assist the command narrative by monitoring the information environment, providing intelligence for specified audiences or individuals, and helping to assess the effects of a narrative.

Intelligence professionals are task-saturated with requirements, so unless a commander makes the collection and analysis of a specific audience a priority, it will likely not get done. Because planners will rarely have the necessary insight into an audience to assist in crafting an effective narrative, they will have to rely on intelligence professionals for support. To achieve this support, planners will have to task intelligence professionals with specific requests to identify behavioral, and cultural attributes. If this information is available, it will often reside in dedicated cells within the J-2, but more often than not will have to be leveraged through reach back support to found within the greater intelligence community. Additionally, the type of information necessary may require an intelligence analyst to fuse multiple types of intelligence to assist (McGuinn, 2019, p. 20). These types of intelligence most often will include open source intelligence (OSINT), human intelligence (HUMINT), and signals intelligence (SIGINT). If done properly, the analyst can provide insight into audiences living in a specific region or locale, relevant actors, and adversaries.

In particular, OSINT can provide significant insight into a population by monitoring social media platforms, blogs, chat rooms, and other sites, which is why it is a growing capability within the Intelligence Community. Specifically, social media monitoring tools allow for the collection and analysis of a large number of selected audiences. To do this type of collection and analysis, various OSINT cells have emerged within the GCCs and larger Intelligence Community. Not all OSINT cells are organized the same, but some have combined OSINT capabilities with their all source analyst cell to provide tailorable intelligence. These cells can create a multitude of products, include "baseball cards" on possible targets, electronic target folders, and key leader assessments.[77]

To be effective, though, there needs to be clear and open lines of communication between the communications and intelligence professionals on staff. In one command, the PAO often went directly to the open source cell with questions and to get a better understanding of what was being said across the information environment about a specific topic. At this command, this was possible because of a strong relationship between the PAO and open source cell directors, and not because it was necessarily codified in a process. The flow of information went both ways though, and the J-2 was kept apprised of PA requests via multiple venues, including the Security Cooperation Working Group (SCWG), IOWG, and direct communication.[78]

---

[77]   Group semistructured interview with experts in Joint Intelligence, in person, February 28, 2020.

[78]   Group semistructured interview with experts in Joint Intelligence, in person, February 28, 2020.

Overcoming Military Cultural Obstacles to Effective Communications

Military culture is dynamic. It can be service-, organization-, and occupation-specific, and it can be a driving force for or against mission accomplishment. Through the course of this research, many interviewees either explicitly or implicitly wanted to know how and why command narrative was important to them. In a command where time and resources are limited, narrative can have a unifying effect on the staff, but that is only if people buy in to the concept. In a very hierarchical organization, such as a GCC staff, rank, position, and culture matter a great deal. When there are cultural barriers between staff functions, it can lead to narrative being undermined. In this section, we discuss some of those divides.

One cultural divide exists between active duty personnel and reservists. Active duty personnel are initially trained in their occupation field, and it isn't until later in their careers that they are afforded broadening opportunities or training. For the influence field, few receive training to think about brand image, marketing, or public relations. [79] Army personnel, in particular, can be very focused on "seizing the objective" rather than more subtle, influence-based activities. Reservists, on the other hand, can bring additional civilian-based expertise to a problem. Not only can this bring a different outlook, but reservists can leverage their experience gleaned from civilian employment.

Another cultural divide can exist between the information professional and combat arms communities. In the latter, there can be scorn or contempt for an idea of command narrative. This all stems from a fundamental misperception of what the narrative is and what it can do for a command. As stated in one interview, "Combat arms says they'll know the truth when they hear it from the barrels of our guns." This goes to show there is a perception in some communities that narrative does nothing for a command; in effect, there is no benefit to having it.

While there are instances of strong coordination and collaboration between information and intelligence professionals, this is not always the case. All too frequently, there is a divide between these two communities that challenges interactions and makes it seem like the PA, Information Operations, and Intelligence sections are working at cross purposes (Schwille et al., 2020). Within the communication capabilities, such as PA, Psychological Operations, and COMSTRAT, this divide can be stark indeed. As one general officer noted, commanders want to have synergy across staff sections and "not have a bunch of stovepipes" (Freedberg, 2019, p. 5). When the cultural differences between Psychological Operations and PA practitioners are allowed to create barriers to coordination and synchronization, it often detracts from a command's ability to generate and present a coordinated message (or narrative). To avoid this, the PA and Psychological Operations personnel need to work together building

---

[79] Semistructured interview conducted with experts in Joint Operations, in person, February 27, 2020.

plans, deconflicting messages in real time so that they do not inadvertently undermine each other (Freedberg, 2019, p. 5).

Even among communications professionals, there can be disagreement over the importance of narrative. One PA expert called command narrative "recreational communication," implying that only communication amateurs are involved.[80] This person went on to say that command narrative, and the processes that enable its creation, such as the CCSWG, are like a heavy sled that has to be dragged around by the command, and that it gives no tangible benefit.

To ensure that command narrative is important to a command, leadership must be involved. Without substantial direction and engagement by a command's leadership for a command narrative, military culture will ensure that narrative does not gain any traction. As one saying goes, "Culture eats strategy for breakfast" (Coffman and Sorensen, 2013). As a command narrative is an enabling portion of a command's strategy, this quote is particularly relevant.

Offensive-Defensive Risk Acceptance

This section deals with commander's risk-versus-reward calculus when it comes to external communications, which can include command narrative. Military commanders are generally risk-adverse in the communications space. This proclivity is reinforced through education and experience. Oftentimes, this risk aversion is the result of operational security measures and the desire to tightly control the image a command projects. However, in recognition of how U.S. adversaries are operating and the perceived success they are reaping from effective operations in the information environment, DoD is slowly coming to the realization that it needs to be more aggressive in the communication space. As one interviewee noted, "DoD is very uncomfortable at describing our role during peacetime, especially in the information environment."[81] One of the best ways for a command to counter adversaries activities in the information environment is to tell its own story. Accomplishing this requires more acceptance of an active, risk-accepting posture in the information environment.

As a product of the military education system, and through cultural learning, commanders tend not to take unnecessary risks. The communication space is viewed to be one in which the wrong tweet, post, or utterance can go dramatically wrong in a hurry. A talking point that is taken out of context or an erroneous tweet can have wide-ranging complications. To prevent this, many commanders hold release authority for external communications to a relatively high level. This more centralized method of control ensures that communications are closely monitored and vetted before being

---

[80]   Semistructured interview conducted with PA, in person, December 23, 2019.

[81]   Interview with PA expert, via phone, March 20, 2020.

released. However, it has the effect of slowing down responses, as each layer of approval takes time. This is not to say that quick responses aren't possible, just that careful review of statements and press releases takes time, slowing down the response process. To avoid this slowdown, some organizations within DoD have developed shortcuts, established standard operating procedures, and pushed authorization for message approval down to lower command echelons.

An example of two organizations that are offensively minded are the Joint Military Information Support Operations (MISO) WebOPS Center (JMWC) and the Information Warfare Task Force–Afghanistan (IWTF-A). Both organizations have well-established guidance on what themes can and cannot be messaged, and they have delegated the responsibility to engage on social media platforms down to lower echelons (and to lower-level officers and enlisted personnel) and are accepting more risk in the communication space.[82] This has dramatically increased their response times while keeping the risk for an erroneous message being sent on social media to an acceptable level.

At the Joint Staff level, communications professionals have developed a communications playbook that provides general guidance to the GCCs across myriad topics. Products like this can be used to do advanced planning, establish approved themes and messages, and help align command narrative (Office of the Secretary of Defense, Assistant to the Secretary for Public Affairs, 2020).

Effective Implementation of Command Narrative

In Chapter Two, we laid out in detail the results of a systematic review of the scholarly literature around narrative as a cognitive framework. In this chapter, we described in detail our insights on command narrative implementation from SME interviews and site visits to the GCCs. We have synthesized insights from both efforts into a set of key summary findings organized by type, and a set of recommendations aimed primarily at the GCC level, presented in the next chapter.

---

[82] Semistructured interview with experts in Joint Operations, in person, February 27, 2020.

CHAPTER FOUR

Conclusion: Effective Command Narrative

This report lays out ways to improve command narrative in U.S. military contexts and is the result of two parallel research efforts. The first was a scholarly one that focused on the scientific literature around narrative in terms of communication and persuasion. The second research effort focused on actual narrative practice in the force, in particular at the GCCs. We were able to offer useful definitions of narrative and related concepts, and we were able to describe how varied and (mis)understood command narrative is in the joint force. These lines of effort produced a set of insights and recommendations for improving the use of narrative—stories as a framing device for making sense of the world. In particular, we laid out specifics on the production, distribution, and assessment of command narrative, with a focus on theater competition for the GCCs. We present these below.

Key Findings

Our SME interviews and site visits to the GCCs revealed many challenges to implementing command narrative. The majority of GCCs do not have command narratives, staff generally do not know what command narrative is, and existing processes and structures generally do not support the development, dissemination, and assessment of effective command narrative. Beyond command narrative, we found a very wide variety of practices supporting communication efforts broadly, many of which seem to hinder effective command narrative.

However, we also found good models for command narrative use, and many important best practices for communication efforts. Below, we have synthesized both our findings from the scholarly literature and practitioner interviews into a consolidated set of key findings and recommendations that follow from those findings.

Narrative matters because it is the primary cognitive framework that human beings use to make sense of the world

Given how fundamental narrative is to human cognition and decisionmaking, communicators must understand the following:

- *Stories play an essential role in how people process information, and effective stories contain common core components.* The literature suggests that that effective stories use characters, traits, goals, motives, conflict and problems, risk and danger, struggles, and details to affect audiences. The best stories include all eight of these components.
- *Narratives are powerful because they influence audiences.* Achieving this type of audience response is also linked to transportation, engagement, memory, and other outcomes. Stories can stir emotions in a way that information and facts cannot.
- *Defeating hostile narratives must go hand in hand with the promotion of alternative, positive narratives.* Audiences will find a narrative or narrative frame for events, and they will make sense of them, one way or another. It is impossible to defeat a narrative and just leave a narrative vacuum. "The one thing that replaces [or modifies] a story-based belief . . . is a better story" (Seese and Haven, 2015, p. 34).
- *Because it is hard to counter an accepted and engaging narrative, it is important to get the command's version of events out first.* Agility and speed are likely to beat mass in terms of narrative.
- *Actions are messages.* Because actions speak louder than words and because there are often preexisting narratives likely to attach to certain actions, commands will need to ensure that deeds and words match. Further, in order to change an existing narrative, commands may need to take new or different actions to create a new, better narrative—narrative is not just a matter of words.
- *Joint doctrine requires the GCC develop a narrative for military operations, one that is synchronized with the U.S. national strategic narrative.* When done properly, command narrative can act as an anchor point for a command's operations and messaging and help ensure that all members of the command are on the same sheet of music.
- *Command narrative will be a priority to the degree that it is a commander's priority.* Because of resource and attention constraints, as well as existing military cultural dispositions, command narrative requires the commander's prioritization to compete for time and attention with other demands.
- *Communication practices and emphases can be personality-dependent.* Unless a given communication practice (such as the use of command narrative) is firmly institutionalized, it may not survive command/personnel turnover.

Effective narrative development must account for complexity—multiple concepts, moderators, and emotional strategies that make narrative more or less effective

- *Understanding the differences among key concepts, including messages, themes, stories, and narrative.* Narratives are a broader system that unites these elements that foster a shared connection for audience members. We found that many experts in staff functions did not know what command narrative is, others confused it with messaging, and still others conflated it with commander's intent. Failure to understand and differentiate between these terms and concepts could hamper the ability of a communicator to develop and implement effective command narrative.

- *Additional moderators—such as causality, temporality, coherency, character identification and proof, and plausibility—can be used to inform a story.* Evidence suggests that these moderators can increase story success by allowing an audience to better follow and identify with a story.
- *Both positive and negative emotions can be effective and influential.* However, encouraging a positive or negative response may lead to different types of outcomes. Storytellers should determine what type of effect they hope to achieve to select the appropriate approach.

Audience is central to effective narrative

Communicators must implement a range of audience-centered best practices to craft effective narratives:

- *Audience understanding should remain central to narrative development and implementation.* Communicators should engage in audience analysis and segmentation prior to story development. This understanding should inform the development and implementation of communication activities.
- *Other communication best practices, such as targeting, tailoring, and framing, can be leveraged in conjunction with narrative theory.* For example, targeting and tailoring can help further customize stories for audiences. Framing literature additionally suggests instances where gain- or loss-framed messages are more likely to persuade individuals. Effective communication requires fine-grained matches with specific audience's language and cultural practices. Broad, generic communication is easier and cheaper to produce but unlikely to resonate with the many local audiences that a command is trying to engage and persuade.

Effective communication practices, including command narrative, require assessment

Effects-based planning calls for measurement and assessment of effects, in an iterative cycle in which the previous efforts assessed effects feed into the next cycle of planning. We found the following:

- *Assessment demands time and resources but is necessary for determining whether communication efforts are making a difference.* Any command narrative effort must measure the performance of specific tasks, actions, events, or programs that can occur prior to, during, and after a message is delivered. Assessment should be guided by theory and rely on validated outcome measures.

- *Good assessment requires good data practices.* Any plausible assessment effort will require robust data collection and analytics.

There is wide variability in how combatant commands organize for and understand the broader issue of coherency in communications

This is reflected in varying military processes and structures that have a powerful impact on the effectiveness of command narrative:

- *Some commands fully integrate all communications through close collaboration and synchronization and anchors on the command's narrative, whereas others seek to deconflict essentially distinct communication efforts, without attention to command narrative.* The latter approach was fraught with difficulty: lack of engagement, inconsistent participation in planning and coordination, failure to collect data systematically, and disagreement about communication ends and means between staff sections.
- *Intelligence has an important part to play in informing communication efforts.* Intelligence professionals are well positioned to inform target audience analysis but are also likely tasked with many other priorities.
- *There are military cultural obstacles to effective communication efforts.* These can include active-duty versus reserve differences, cultural differences and tensions between combat arms and information practitioners, and even tensions between different information practitioner lines. An example of the latter is certain interpretations of the "PAO firewall," whereby PA staffs maintain willful ignorance of the efforts of other information professionals and refuse to fully synchronize PA with the output of other capabilities.
- *There is no clear staff section or office at the GCC level responsible for narrative development.* Having a "first among equals" tasked as primarily responsible for narrative development could provide a needed forcing function.

Communication efforts involve risk, and thus risk-aversion may hinder communication

Because communication errors can be so public and so damaging, there are strong incentives to not take risks. At the same time, there are few direct rewards for effective communication.

Recommendations

GCCs should consider how to leverage insights gleaned from our analysis. Taken individually or used in conjunction with one another, these recommendations can improve the development, implementation, or evaluation of narrative activities (including work related to messages, themes, or stories). While these are primarily framed around the GCCs, they could be adapted down to the component level, or up to multinational force levels (i.e., NATO). Drawing on our analysis of both lessons learned from interviews and site visits and building on the broader insights from our scientific literature around narrative and communication, we offer the following recommendations.

Systematically implement an evidence-based command narrative that supports the command's campaign objectives

- *Use narrative as a strategy to inform, engage, model behavior, persuade, and provide comfort* (Shaffer et al., 2018, p. 434). Evidence suggests that narrative communication can be more effective than other forms of communication approaches that rely on facts, logic, and/or statistics alone.
- *Be prepared for dynamic change in using a narrative strategy.* Narrative "in the wild" is dynamic and both changes and spreads. Communicators should treat audiences as both coauthors and recipients of any narrative (Zalman, 2010, p. 6).
- *Leverage everyone in the command's potential to contribute to (or undermine) intended narratives or stories.* Make sure that everyone in the command, regardless of rank or billet, has enough guidance for and understanding of the intended narrative.
- *Make command narrative and communication a priority.* There are many obstacles to effective communication and command narrative use: for example, lack of conceptual clarity in the force, assessment difficulty, cultural gaps between military functions, risk aversion, and intelligence focus on other kinds of targets. Unless the commander makes effective communication and use of command narrative a priority, they are unlikely to happen.

Generate and maintain trust with audiences

It is easy to say or do the wrong thing and lose credibility, undermining the effectiveness of command narrative.

- *Avoid the say-do gap.* Make sure that narratives that include accounts of friendly actions are consistent with those actions, and vice versa. Be ready to change planned actions to be consistent with more favorable narratives.
- *If using fictional elements in a narrative, make sure that is done clearly and intentionally.* Lack of clarity related to fictional versus nonfictional elements can lead to diminished credibility or assumptions related to deception.

Put the audience—not the command—at the center of narrative crafting

While narrative must serve the command's campaign goals, effective narrative centers on the intended audiences:

- *Identify and research existing narratives in the area of responsibility.* What are the long-standing knowledge, attitudes, and beliefs that audiences widely accept? What stories shape an audience's perceptions and worldview? These details are critical to crafting an alternative story that will resonate with intended audiences.
- *Conduct formative audience research.* Along with narrative understanding, audience analysis provides additional understanding related to intended markets. What are their preferred media channels and sources? What opportunities exist to meaningfully engage with intended audiences and what challenges can be expected? If limited opportunities existing for primary research, secondary research involving existing data can still provide some perspective on audience members.
- *Use audience segmentation to distinguish groups with different characteristics.* Be thoughtful about what is desired from each segment and how each segment might interact with the narrative. Some segments will be more amenable to changing attitudes and behaviors than others.

Implement best practices in crafting narratives

- *GCCs should nest narrative development from top to bottom.* Ensure that stories, themes, and messages align and support the desired narrative.
- *Use all eight story components when writing a story (to support a narrative).* Missing story elements will make a story less powerful. When possible, use characters the audience will identify with.
- *Create stories that transport individuals, where the audience is "caught up in the story."* Feelings of transportation are often linked to other desired outcomes.
- *Use additional moderators—such as causality, temporality, coherency, character identification and proof, and plausibility—based on desired story goals and objectives.* Moderators are optional but potentially helpful strategies for crafting effective stories.
- *Use additional best practices, such as positive or negative emotions, targeting, tailoring, and framing.* Using the best practices are optional, but also important tools for increasing story success.
- *Use assessment to make sure communication is making a difference.* Put assessment plans in place and ensure dedicated data collection to support assessment.

Implement best communication practices in military processes and structures

- *GCCs should tear down the PAO firewall.* PAO can best contribute to theater competition through close collaboration with other information capabilities and an integration mindset. PAO is constrained to wholly truthful communication and needs to protect its credibility, but must have situational awareness of other communication efforts and participate in and support vigorous communication synchronization processes across the command.
- *GCCs should adopt a communications integration mindset and a close collaboration model.* Post-hoc deconfliction between separate, stove-piped communication efforts is not in accord with doctrine and is unlikely to be effective.
- *GCCs should put out annual training guidance on narrative.* The entire command should understand the command narrative as it is the anchor for deeds and words, as well as foundational concepts and vocabulary. The entire command—including junior personnel—should understand the command's narrative and how it can help guide their decisions. In February 2020, the office of the Assistant to the Secretary for Public Affairs published the *DoD Communication Playbook*. This document can serve as a template for what each GCC could produce to help align down trace organizations to the command's themes and narrative.
- *GCCs should build robust communication assessment capabilities.* Such capability requires identifying smart objectives, collecting and analyzing relevant data, leveraging data storage and analytics systems, and producing outputs that turn findings into actionable insight that can inform the next round of communication planning.
- *Use effects-based planning in narrative development.* Like all operations, effective narrative by identifying the desired effects and planning backwards to achieve those effects.
- *Prioritize commander involvement. Effective use of command narrative requires attention and emphasis from the commander.* Commanders should hold leaders responsible for close collaboration on narrative, be involved in narrative development, and either the commander or their deputy should be involved in ongoing communication synchronization processes.
- *GCCs should invest in narrative development. Effective narrative that produces effects will require a right-sized force.* For example, tailoring messages to the local culture and language variation of a target audience is an important factor in a successful narrative, and requires particular expertise and capacity.
- *GCCs should designate a "keeper of the narrative" tasked with primary responsibility for narrative development.* While the specific staff section or entity for this may vary between commands, having a first among equals for narrative development will provide a needed forcing function for narrative.

- *Narrative processes and structures must become institutionalized.* For narrative in particular and communication broadly, processes and structures need to be baked into the institution, as personality-driven efforts will not suffice as personnel rotate out of the command.

Our recommendations involve substantive, potentially disruptive changes. For example, we found that the common implementation of a firewall between PAO and the other staff informational activities is both deeply entrenched and likely hinders GCCs' ability to compete. Tearing that firewall down is a nontrivial change to make and will require local, contextualized work at each GCC. Similarly, asking GCCs to collect structured data across a range of campaign-related objectives and to apply analytics to that data to perform robust assessment is necessary, but not an easy change. It will require changes in practice and investment in resources.

These changes are necessary though if GCCs wish to compete agilely and effectively. As we note repeatedly throughout this report, everything commands do (and do not do) sends powerful messages that multiple audiences interpret using narrative frameworks. If GCCs and other commands with operational and strategic concerns wish to have campaign-relevant effects, they will need to be able to communicate in a coherent, effective way that leverage humans' storytelling nature, and thus anchor words and deeds in a harmonious way that enables the commander's intent.

Interview Protocol and Read-Ahead

Command Narrative Interview Read-Ahead

This interview read-ahead is in support of the RAND Corporation project, "Identifying organizational structure and processes for the development, integration, and assessment of command narrative."

The project is focused on three areas, which relate to narrative formation, projection, and reception:

- **Formation:** What processes should the command use in developing its command narrative? That is, how should they decide what the command wants to be communicating, and identifying things like supporting themes, messages, and activities?
- **Projection:** Once a narrative is conceived/designed/planned, how should the command propagate, integrate, coordinate, and deconflict actions and communications to support that narrative across the command?
- **Reception:** How should the command assess the different aspects of the command narrative process (i.e., to include the planning and deconfliction of narrative, but also the effectiveness of the narrative)?

Background and Overview

1. Please tell us about your office/organization and its responsibilities.
2. What is your position here, and how does it relate to command narrative or communication? Please let us know if you have prior experience/postings related to this topic.

General Questions

1. How does Public Affairs integrate with other communication/ informational efforts.

    a. What are the processes they use?

    b. Specifically, what does integration with Military Information Support Operations (MISO)/Psychological Operations efforts?

2. How does Communication synchronization and deconfliction take place the organization?

    a. What are the organizational structures (e.g., working groups, divisions/functional areas)?

    b. What are the organizational processes (database entry, products/ spreadsheets, trackers)?

3. What are the attitudes and understanding around command narrative?

    a. What is "Narrative" conceptually?

    b. How would you define "Command narrative"?

    c. How is command narrative distinct from commander's intent, or the commander's priorities?

4. What are the organizational structures that support production, distribution, and assessment of narrative?

5. What are the organizational process that support production, distribution, and assessment of narrative?

Catch All

1. Before we end, is there anything else you'd like to add?
2. Is there anyone else you recommend that we talk to?

Handout

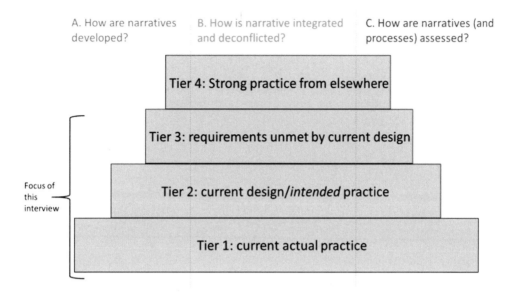

Methods for Literature Review

Identifying Articles for Review

This research effort included of a search of peer-reviewed literature to identify best practices and the empirically defined characteristics and qualities of effective narratives. The search focused on journals published in English from January 2000 to October 2019 (the month the search was conducted).[1] Six databases were utilized in the search: Academic Search Complete, PsycINFO, Scopus, Web of Science, ProQuest Military Database, and the Defense Technical Information Center. The selection of these databases ensured that the team reviewed both traditional peer-reviewed academic literature and practitioner-based literature. In searching these databases, sources were identified using a keyword strategy based in five areas of study as it relates to the analysis of narratives: (1) keywords specific to communication and narratives, (2) keywords specific to linguistics, (3) keywords specific to psychology, (4) keywords specific to neuroscience, and (5) keywords specific to defense studies. These keywords were selected through a series of test searches to determine which keywords identified the most relevant literature. Each search was conducted by combining a keyword from communications (10 in total) with a keyword from one of the other four categories (19 in total).

Table B.1 includes the search strings utilized for the literature search. In an effort to focus the results on literature related to the effectiveness and evaluation of narratives, the terms *effect** and *evaluat** were also added at the end of each search. Both of these two terms were stemmed to capture all variations, such as *effect* and *effective* as well as *evaluate* and *evaluation*. In the next paragraph, the search strategy is outlined.

In an effort to narrow the list of search results and help ensure the greatest number of relevant sources, the following strategy was applied to the database queries. For keywords from the communications column, the search was limited strictly to instances

---

[1] Our goal was to produce a large but manageable amount of literature to include in our search. While research earlier than 2000 might be relevant, we bounded our search there because the communication landscape has changed dramatically over the past 20 years, particularly as it relates to strategic communication, information campaigns, and social marketing.

Table B.1
Search Strings for the Literature Search Process

Communications	Linguistics	Psychology	Neuroscience	Defense-Specific
Narrative	Sociolinguistics or Socio-linguistics	Behavior or Behaviour	"Emotional Arousal"	"Information Operations"
"Strategic Narrative"	"Discourse Analysis"	"Behavior Change" or "Behaviour Change"	Empathy	"Command Narrative"
"Effective Narrative"	"Corpus Linguistics"	Influence	"Social Neuroscience" or "Social-Neuroscience"	Propaganda
Story	"Content Analysis"	Persuade		Terrorism
Storytelling		Attitude		
"Strategic Communication"		"Cognitive Process"		
"Information Campaign"				
"Social Marketing"				
"Media Campaign"				
Message				

NOTE: In addition to these, the terms the terms *effect** and *evaluat** were also added at the end of each search.

in which the term appeared in the title of the publication. For keywords from the other four columns, the search was limited to instances in which the term appeared in the abstract of the publication. However, after initial testing in several of the selected databases, it was determined that when certain keywords from the communication column [*narrative, story, message*] were combined with certain keywords from the other columns [*behavior, influence, attitude, empathy, engagement*], it produced a significantly high number of results with many not lending useful information to this study. As a result, when conducting a search using one of these three communication terms [*narrative, story, message*], the search strategy was slightly altered and split into two ways.

When conducting a search for most keywords from the other four columns [*sociolinguistics, discourse analysis, behavior change*, etc.], the strategy remained the same, where the communications term was limited to the article title and the terms from the other four columns were limited to appearances in the abstract. But when searching for one of the five identified keywords [*behavior, influence, attitude, empathy, engagement*], the queries were limited to instances in which all keywords appeared in the title of the source. For terms *effect** and *evaluat** that are ended at the end or the search, these always remain limited to the article's abstract. Below is an example of this strategy when searching for publications related to *narrative*.

- **Typical search strategy:** "narrative" AND ("socio-linguistics" OR sociolinguistics OR "discourse analysis" OR "corpus linguistics" OR "behavior change" OR "behaviour change" OR persuade OR "cognitive process" OR "emotional arousal" OR "social neuroscience" OR "social neuro-science" OR "information operations" OR "command narrative" OR propaganda OR terrorism) AND (effect* OR evaluat*)
  - The communications term *narrative* is limited strictly to appearances in the article "title," and the remaining terms are limited to the article's "abstract."
- **Altered search strategy:** "narrative" AND (behavior OR behaviour OR influence OR attitude OR empathy OR engagement)
  - For all terms, the results are limited to appearances in the article's "title." However, the terms *effect** and *evaluat** continue to be searched within the article's "abstract."

The initial search produced 5,197 results, with the results for each database illustrated in Figure B.1. After the removal of duplicate citations and 13 citations from the Defense Technical Information Center that we did not have access to, we had a total of 2,174 unique articles to review.

Title, Abstract, and Full-Text Review

Because of the high volume of results, we constructed a labeling system of four categories to determine which articles were most crucial to review for the study. This process of categorizing articles was conducted by reading over the abstract for each of the 2,174 articles and binning them into one of four categories. These categories are *Crucial*: articles that the study will focus on, *Highly Recommend*: articles that are still relevant but will not be reviewed in the interests of time, *If Time Allows*: articles that have relevant content but are not as pertinent to the study, and *Delete*: articles that were included in the dataset by chance. To help ensure accurate categorization, a second team member reviewed several citations labeled as crucial to provide intercoder reliability. These categories, along with their corresponding criteria are listed in Table B.2. Table B.3 provides an exemplar article for each category and a brief explanation for the categorization.

The categorization process allowed us to go from 2,174 articles to 101 *crucial* articles to comprehensively review for the study. When reviewing the articles, there were four articles that the team was unable to access without paying for the articles individually, which brought the new total to 97 articles. After conducting a full review for

Figure B.1
Initial Database Results

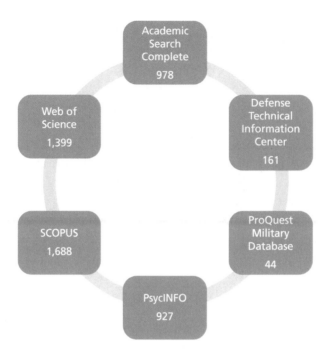

each of the 97 articles, we determined that seven of articles did not meet the criteria we set for crucial sources and, as such, excluded these sources from inclusion in the report. This brings the total number of crucial articles reviewed from the literature review to 90 articles. The final breakdown for each category is listed below.

- Crucial: 90
- Highly Recommend: 474
- If Time Allows: 1,510
- Delete: 96.

To supplement this list, approximately 15 additional sources were collected from citations found in the articles we reviewed and other sources the team identified as relevant to the study.

Table B.2
Categorization Table

Category	Methods	Rhetor	Topic/Content	Other
Crucial	Employed methods provide a comprehensive/generalizable overview about narrative (examples: meta-analysis, literature reviews, large survey)	The rhetor is the military or U.S. government (including U.S. Department of Health and Human Services entities). Individual military actors should be considered crucial	The article *heavily focuses* on defining narrative, describing the components of narrative, best practices, or provides specific guidance for the military	EUCOM area of responsibility
Highly Recommend	Employed methods provide an adequate overview about narrative (examples: original study with a literature review section, case study, single experiment)	Rhetor is any other organization (examples: nonprofit entities, for-profit businesses, community based organizations, faith-based organizations)	The article *might focus* on defining narrative, describing the components of narrative, best practices, or provides specific guidance for the military	Outside EUCOM area of responsibility
If Time Allows	Employed methods provide an insufficient overview about narrative (examples: article contains no literature review; methods are not described or poorly executed)	Rhetor is an individual (example: Queen Elizabeth)	The article *does not* define narrative, describing the components of narrative, best practices, or provides specific guidance for the military	Not applicable: Study does not define a geographical descriptor
Other directions for moving articles into "If Time Allows" category	• Articles about students (K–12)/adolescents • Articles about cognitive disabilities • Articles that perform a content analysis (example: newspapers) and/or perform a thematic analysis of a topic			
Flag	Article that needs more attention or additional review			
Irrelevant	Included in returned dataset by chance (the article is clearly included in sample by chance/accident)			

Table B.3
Categorization Exemplar Articles

Category	Title	Explanation
Crucial	Meta-Analytic Evidence for the Persuasive Effect of Narratives on Beliefs, Attitudes, Intentions, and Behaviors (Braddock and Dillard, 2016)	This study is a meta-analysis on studies that evaluated narrative's persuasive influence and seeks to provide a more precise test of a narrative's effect on beliefs, attitudes, intentions, and behaviors.
Highly Recommend	Why Story Matters: A Review of Narrative in Serious Games (Naul and Liu, 2020)	While the study identifies different narrative elements that have shown to be effective, the article does not focus on changing audience behavior or attitude and does not provide clear guidance for military or government.
If Time Allows	Children's Memory for Narratives: Influence of Content Familiarity and Input Modality (Turella et al., 2002)	The study focused on story recall, but the subjects of the study were eight-year-old children enrolled in elementary school, which we have elected to move to this category given our study's focus on adults rather than adolescents.
Delete	Evaluation of the Influence of Vertical Irregularities on the Seismic Performance of a Nine-Story Steel Frame (Michalis, Dimitrios, and Manolis, 2006)	Article included in the dataset by chance as the focus is strictly on physical building structure.

References

Aguilar, Kirsten G., *Public Affairs: A Strategic Communication Force Multiplier*, Newport, R.I.: Naval War College, thesis, May 2011.

Ajzen, Icek, "The Theory of Planned Behaviour: Reactions and Reflections," *Psychology & Health*, Vol. 26, No. 9, 2011.

Baldwin, Robert F., *A New Military Strategic Communications System*, Fort Leavenworth, Kan.: Command and General Staff College, May 2007.

Ballard, Andrew O., D. Sunshine Hillygus, and Tobias Konitzer, "Campaigning Online: Web Display Ads in the 2012 Presidential Campaign," *PS: Political Science & Politics*, Vol. 49, No. 3, July 2016, pp. 414–419.

Berger, Charles R., Yerheen Ha, Meng Chen, "Story Appraisal Theory: From Story Kernel Appraisals to Implications and Impact," *Communication Research*, Vol., 46 No. 3, 2016, pp. 303–332.

Bernardi, Daniel L., Pauline Cheong, and Chris Lundry, *Narrative Landmines: Rumors, Islamist Extremism, and the Struggle for Strategic Influence*, New Brunswick, N.J.: Rutgers University Press, 2012.

Boerman, Sophie C., Sanne Kruikemeier, and Frederik J. Zuiderveen Borgesius, "Online Behavioral Advertising: A Literature Review and Research Agenda," *Journal of Advertising*, Vol. 46, No. 3, 2017, pp. 363–376.

Braddock, Kurt, "The Utility of Narratives for Promoting Radicalization: The Case of the Animal Liberation Front," *Dynamics of Asymmetric Conflict*, Vol 8, No. 1, 2015, pp. 38–59.

Braddock, Kurt, and James Price Dillard, "Meta-Analytic Evidence for the Persuasive Effect of Narratives on Beliefs, Attitudes, Intentions, And Behaviors," *Communication Monographs*, Vol. 83, No. 4, 2016, pp. 446–467.

Case, Dean J., II, and Brian C. Mellen, *Changing the Story: The Role of the Narrative in the Success or Failure of Terrorist Groups*, Monterey, Calif.: Naval Postgraduate School, thesis, December 2009.

Casebeer, William D., and James A. Russell, "Storytelling and Terrorism: Towards a Comprehensive 'Counter-Narrative Strategy,'" *Strategic Insights*, Vol. 4, No. 3, March 2005.

Cheng, Tania, Danielle K. Woon, and Jennifer K. Lynes, "The Use of Message Framing in the Promotion of Environmentally Sustainable Behaviors," *Social Marketing Quarterly*, Vol. 17, No. 2, 2011, pp. 48–62.

Churchill, Susan, Louisa Pavey, Donna Jessop, and Paul Sparks, "Persuading People to Drink Less Alcohol: The Role of Message Framing, Temporal Focus and Autonomy," *Alcohol and Alcoholism*, Vol. 51, No. 6, 2016, pp. 727–733.

Coffman, Curt, and Kathie Sorensen, *Culture Eats Strategy for Lunch: The Secret of Extraordinary Results, Igniting the Passion Within*, Denver, Colo.: Liang Addison Press, September 2013.

Corman, Steven R., "Understanding Extremists' Use of Narrative to Influence Contest Populations," position paper submitted to Workshop on Mapping Ideas: Discovering and Information Landscapes, San Diego State University, June 29–30, 2011.

Davidson, Brett, "Storytelling and Evidence-Based Policy: Lessons from the Grey Literature," *Palgrave Communications*, September 12, 2017.

de Graaf, Anneke, "The Effectiveness of Adaptation of the Protagonist in Narrative Impact: Similarity Influences Health Beliefs Through Self-Referencing," *Human Communication Research*, Vol. 40, No. 1, January 2014, pp. 73–90.

Dillard, James, and Nabi, Robin, "The Persuasive Influence of Emotion in Cancer Prevention and Detection Messages," *Journal of Communication*, Vol. 56, August 2006, pp. S123–S139.

Dillard, James Price, Lijiang Shen, and Renata Grillova Vail, "Does Perceived Message Effectiveness Cause Persuasion or Vice Versa? 17 Consistent Answers," *Human Communication Research*, Vol. 33, No. 4, 2007.

DoD—*See* U.S. Department of Defense.

Estle, Sara J., Leonard Green, Joel Myerson, and Daniel D. Holt, "Differential Effects of Amount on Temporal and Probability Discounting of Gains and Losses," *Memory & Cognition*, Vol. 34, 2006, pp. 914–928.

Forman, Janice, *Storytelling In Business: The Authentic and Fluent Organization*, Redwood City, Calif.: Stanford University Press, 2013.

Freedberg Sydney J., Jr., "'Desperate Need for Speed' as Army Takes on Chinese, Russian, ISIS Info Ops," *Breaking Defense*, August 21, 2019.

French, Jeff, "Using Social Marketing to Improve Preparedness for Pandemics: The Work of the Ecom Program," *Social Marketing Quarterly*, Vol. 22, No. 2, 2016, pp. 138–142.

Gallagher, Kristel M., and John A Updegraff, "Health Message Framing Effects on Attitudes, Intentions, And Behavior: A Meta-Analytic Review," *Annals of Behavioral Medicine*, Vol. 43, No. 1, February 2012, pp. 101–116.

Gerend, Mary A., and Margaret Cullen, "Effects of Message Framing and Temporal Context on College Student Drinking Behavior," *Journal of Experimental Social Psychology*, Vol. 44, No. 4, July 2008, pp. 1167–1173.

Gianfranco Walsh, Louise M. Hassan, Edward Shiu, J. Craig Andrews, and Gerard Hastings, "Segmentation in Social Marketing: Insights from The European Union's Multi-Country, Antismoking Campaign," *European Journal of Marketing*, Vol. 44, No. 7/8, 2010.

Green, Melanie C., and Timothy C. Brock, "The Role of Transportation in the Persuasiveness of Public Narratives," *Journal of Personality and Social Psychology*, Vol. 79, No. 5, November 2000, pp. 701–721.

Green, Melanie C., and Jenna L. Clark, "Transportation into Narrative Worlds: Implications for Entertainment Media Influences on Tobacco Use," *Addiction*, Vol. 108, No. 3, March 2013, pp. 477–484.

Hamby, Anne, Ulrich Ecker, and David Brinberg, "How Stories in Memory Perpetuate the Continued Influence of False Information," *Journal of Consumer Psychology*, Vol. 30, No. 2, April 2020, pp. 240–259.

Haven, Kendall, *Story Smart: Using the Science of Story to Persuade, Influence, Inspire, and Teach*, Santa Barbara, Calif.: Libraries Unlimited, 2014.

Hester, Casey, and Ronald Schleifer, "Enhancing Physician Empathy: Optimizing Learner Potential for Narrative Transportation," *Enthymema-International Journal of Literary Criticism Literary Theory and Philosophy of Literature*, No. 16, 2016, pp. 105–109.

Hinyard, Leslie. J., and Matthew W. Kreuter, "Using Narrative Communication as A Tool for Health Behavior Change: A Conceptual, Theoretical, and Empirical Overview," *Health Education and Behavior*, Vol. 34, No. 5, 2007.

Hochman, Yael, and Gabriela Spector-Mersel, "Three Strategies for Doing Narrative Resistance: Navigating Between Master Narratives," *British Journal of Psychology*, Vol. 59, No. 4, October 2020, pp. 1043–1061.

Iqbal, Khuram, Saad Kalim Zafar, and Zahid Mehmood, "Critical Evaluation of Pakistan's Counter-Narrative Efforts," *Journal of Policing, Intelligence and Counter Terrorism*, Vol. 14, No. 2, 2019, pp. 147–163.

Jin, Seunga Venus, Joe Phua, Kwan Min Lee, "Telling Stories About Breastfeeding Through Facebook: The Impact of User-Generated Content (UGC) On Pro-Breastfeeding Attitudes," *Computers in Human Behavior*, Vol. 46, May 2015, pp. 6–17.

Johnson, David G., *Strategic Communication: The Meaning is in the People*, Carlisle, Pa.: U.S. Army War College, thesis, June 2011.

Joint Doctrine Note 2-13, *Commander's Communication Synchronization*, Washington D.C.: U.S. Department of Defense, 2013.

Joint Publication 1, *Doctrine for the Armed Forces of the United States*, Washington, D.C.: U.S. Joint Chiefs of Staff, March 25, 2013, incorporating change 1, July 12, 2017.

Joint Publication 2-01.3, *Joint Intelligence Preparation of the Operating Environment*, Washington, D.C.: U.S. Joint Chiefs of Staff, May 21, 2014.

Joint Publication 3-0, *Joint Operations*, Washington, D.C.: U.S. Joint Chiefs of Staff, January 17, 2017, incorporating change 1, October 22, 2018.

Joint Publication 3-61, *Public Affairs*, Washington, D.C.: U.S. Joint Chiefs of Staff, November 7, 2015, incorporating change 1, August 19, 2016.

Joint Publication 5-0, *Joint Planning*, Washington, D.C.: U.S. Joint Chiefs of Staff, June 16, 2017.

Joint Staff J-7, Deployable Training Division, *Communication Strategy and Synchronization*, Insights and Best Practices Focus Paper, May 2016.

Joint Staff J-7, Joint and Coalition Warfighting, *Commander's Handbook for Assessment Planning and Execution: Version 1.0*, Suffolk, Va., 2011.

JP—*See* Joint Publication.

Kahneman, Daniel, and Amos Tversky, "Prospect Theory: An Analysis of Decision Under Risk," *Econometrica*, Vol. 47, 1979.

Key, Thomas Martin, and Andrew J. Czaplewski, "Upstream Social Marketing Strategy: An Integrated Marketing Communications Approach," *Business Horizons*, Vol. 60, No. 3, 2017, pp. 325–333.

Kim, Kenneth E., "Framing as a Strategic Persuasive Message Tactic," in Derina Holtzhausen and Ansgar Zerfass, eds., *The Routledge Handbook of Strategic Communication*, Abingdon, UK: Routledge, 2014.

Kubacki, Krzysztof, Sharyn Rundle-Thiele, Bo Pang, and Nuray Buyucek, "Minimizing Alcohol Harm: A Systematic Social Marketing Review (2000–2014)," *Journal of Business Research*, Vol. 68, No. 10, October 2015, pp. 2214–2222.

Laity, Mark, "NATO and the Power of Narrative," *Beyond Propaganda*, September 2015, pp. 22–28.

Lam, Chris, and Mark A. Hannah, "Flipping the Audience Script: An Activity That Integrates Research and Audience Analysis," *Business and Professional Communication Quarterly*, Vol. 79, No. 1, 2016.

LeadSync, *Facebook Audience Targeting Guide: The Comprehensive Targeting Guide*, undated. As of January 21, 2021:
https://leadsync.me/blog/wp-content/uploads/2018/11/Facebook-Ads-Audience-Targeting-Guide-1.pdf

Maan, Ajit, "Narratives Are About 'Meaning,' Not 'Truth,'" *Foreign Policy*, December 3, 2015.

Mah, Manuel W., Yat Cho Tam, and Sameer Deshpande, "Social Marketing Analysis of 20 Years of Hand Hygiene Promotion," *Infection Control and Hospital Epidemiology*, Vol. 29, No. 3, March 2008, pp. 262–270.

Mahood, Samantha, and Halim Rane, "Islamist Narratives in ISIS Recruitment Propaganda," *Journal of International Communication*, Vol. 23, No. 1, 2017.

Marcellino, William, Christopher Paul, Elizabeth L. Petrun Sayers, Michael Schwille, Ryan Bauer, Jason Vick, and Walter F. Landgraf III, *Command Narrative Smart Guide*, Santa Monica, Calif.: RAND Corporation, TL-A353-1, 2021. As of March 31, 2021:
https://www.rand.org/pubs/tools/TLA353-1.html

McGuinn, Phillip, *Joint Force Narrative-Led Operations with Public Affairs as "Keepers of the Narrative"*, Washington, D.C.: National Defense University College of Information and Cyberspace School of Joint Strategic Studies, thesis, May 1, 2019.

Meier, Benjamin D., *The Operational Narrative in Wars of Choice*, Fort Leavenworth, Kan.: U.S. Army Command and General Staff College, 2016.

Metz, Thomas F., Mark W. Garret, James E. Hutton, and Timothy Bush, "Massing Effects in the Information Domain—A Case Study in Aggressive Information Operations," *Military Review*, May–June 2006.

Michalis, Fragiadakis, Vamvatsikos Dimitrios, and Papadrakakis Manolis, "Evaluation of the Influence of Vertical Irregularities on the Seismic Performance of a Nine-story Steel Frame," *Earthquake Engineering and Structural Dynamics*, Vol. 35, 2006, pp. 1489–1509.

Moore, Charles L., Brian Steed, Sohail Shaikh, Dana Eyre, Ian McCulloh, Jason Spitaletta, Randall Munch, and Chris Worret, "SMA White Paper: Maneuver and Engagement in the Narrative Space," Strategic Multilayer Assessment (SMA), white paper, January 2016. As of February 10, 2021:
http://nsiteam.com/social/wp-content/uploads/2015/12/Maneuver-in-the-Narrative-Space_Final_Jan2016.pdf

Morris, Brandi S., Polymeros Chrysochou, Jacob D. Christensen, Jacob L. Orquin, Jorge Barraza, Paul J. Zak, and Panagiotis Mitkidis, "Stories vs. Facts: Triggering Emotion and Action-Taking on Climate Change," *Climatic Change*, Vol. 154, Nos. 1–2, 2019, pp. 19–36.

Multinational Information Operations Experiment, *Narrative Development in Coalition Operations*, Draft V0.96, January 10, 2014.

Murphy, Dennis M., *The Trouble with Strategic Communication(s)*, Carlisle Barracks, Pa.: U.S. Army War College, Center for Strategic Leadership Issue Paper, Vol. 2-08, January 2008.

Murrar, Sohad, and Markus Brauer, "Overcoming Resistance to Change: Using Narratives to Create More Positive Intergroup Attitudes," *Current Directions in Psychological Science*, Vol. 28, No. 2, 2009, pp. 164–169.

Naul, E., and M. Liu, "Why Story Matters: A Review of Narrative in Serious Games," *Journal of Educational Computing Research*, Vol. 58, No. 3, 2020, pp. 687–707.

Nissen, Thomas Elkjer, "Narrative Led Operations: Put the Narrative First," *Small Wars Journal*, October 17, 2012.

Nissen, Thomas Elkjer, "Narrative Led Operations," *Militært Tidsskrift*, No. 141, January 2013, pp. 67–77.

Noar, Seth M., "A 10-Year Retrospective of Research in Health Mass Media Campaigns: Where Do We Go from Here?" *Journal of Health Communication*, Vol. 11, No. 1, 2006.

Noar, Seth M., Nancy Grant Harrington, and Rosalie Shemanski Aldrich, "The Role of Message Tailoring in the Development of Persuasive Health Communication Messages," *Annals of the International Communication Association*, Vol. 33, No. 1, 2009, pp. 73–133.

Office of the Secretary of Defense, Assistant to the Secretary for Public Affairs, *DoD Communication Playbook*, Washington, D.C., February 2020.

Ooms, Joëlle A., Carel J. M. Jansen, and John C. J. Hoeks, "The Story Against Smoking: An Exploratory Study into the Processing and Perceived Effectiveness of Narrative Visual Smoking Warnings," *Health Education Journal*, Vol. 79, No. 2, 2019, pp. 166–179.

Palmer, David C., "The Power of Narratives Derives from Evoked Behavior," *Perspectives on Behavior Science*, Vol. 41, 2018, pp. 503–507.

Paul, Christopher, "Getting Better at Strategic Communication," testimony presented before the House Armed Services Committee, Subcommittee on Emerging Threats and Capabilities, July 12, 2011.

Paul, Christopher, *Assessing and Evaluating Department of Defense Efforts to Inform, Influence, and Persuade: Worked Example*, Santa Monica, Calif.: RAND Corporation, RR-809/4-OSD, 2017. As of January 21, 2021:
https://www.rand.org/pubs/research_reports/RR809z4.html

Paul, Christopher, Kristen S. Colley, and Laura Steckman, "Fighting Against, With, and Through Narrative: Developing the Reasons Why We Are There," *Marine Corps Gazette*, March 2019.

Paul, Christopher, and Miriam Matthews, *The Language of Inform, Influence, and Persuade: Assessment Lexicon and Usage Guide for U.S. European Command Efforts*, Santa Monica, Calif.: RAND Corporation, RR-2655-EUCOM, 2018. As of January 21, 2021:
https://www.rand.org/pubs/research_reports/RR2655.html

Paul, Christopher, Jessica Yeats, Colin P. Clarke, Miriam Matthews, and Lauren Skrabala, *Assessing and Evaluating Department of Defense Efforts to Inform, Influence, and Persuade: Handbook for Practitioners*, Santa Monica, Calif.: RAND Corporation, RR-809/2-OSD, 2015. As of January 21, 2021:
https://www.rand.org/pubs/research_reports/RR809z2.html

Sallis, Anna, Hugo Harper, and Michael Sanders, "Effect of Persuasive Messages on National Health Service Organ Donor Registrations: A Pragmatic Quasi-Randomised Controlled Trial with One Million UK Road Taxpayers," *Trials*, Vol. 19, No. 1, 2018.

Schmid, Kristina L., Susan E. Rivers, Amy E. Latimer, and Peter Salovey, "Targeting or Tailoring? Maximizing Resources to Create Effective Health Communications," *Marketing Health Services*, Vol. 28, No. 1, 2008, pp. 32–37.

Schouten, Dustin J., *U.S. Strategic Communications Against Islamic Fundamentalists*, Monterey, Calif.: Naval Postgraduate School, thesis, March 2016.

Schwille, Michael, Anthony Atler, Jonathan Welch, Christopher Paul, and Richard C. Baffa, *Intelligence Support for Operations in the Information Environment: Dividing Roles and Responsibilities Between Intelligence and Information Professionals*, Santa Monica, Calif.: RAND Corporation, RR-3161-EUCOM, 2020. As of January 21, 2021: https://www.rand.org/pubs/research_reports/RR3161.html

Seese, Gregory S., and Kendall Haven, "The Neuroscience of Influential Strategic Narratives and Storylines," *IO Sphere*, Fall 2015, pp. 33–38.

Shaffer, Victoria A., Elizabeth S. Focella, Andrew Hathaway, Laura D. Scherer, and Brian J. Zikmund-Fisher, "On the Usefulness of Narratives: An Interdisciplinary Review and Theoretical Model," *Annals of Behavioral Medicine*, Vol. 52, No. 5, 2018, pp. 429–442.

Shen, Lijiang, Suyeun Seung, Kristin K. Andersen, and Demetria McNeal, "The Psychological Mechanisms of Persuasive Impact from Narrative Communication," *Studies in Communication Sciences*, Vol. 17, No. 2, 2017, pp. 165–181.

Simpson, Emile, *War from the Ground Up: Twenty-First-Century Combat as Politics*, Oxford, UK: Oxford University Press, 2012.

Slater, Michael D., "Theory and Method in Health Audience Segmentation," *Journal of Health Communication*, Vol. 1, No. 3, 1996.

Stead, Martine, Stephen Tagg, Anne Marie MacKintosh, and Douglas Eadie, "Development and Evaluation of a Mass Media Theory of Planned Behaviour Intervention to Reduce Speeding," *Health Education Research*, Vol. 20, No. 1, 2005, pp. 36–50.

Stephenson, Michael, and Philip Palmgreen, "Sensation Seeking, Perceived Message Sensation Value, Personal Involvement, and Processing of Anti-Marijuana PSAs," *Communication Monographs*, Vol. 68, No. 1, 2001, pp. 49–71.

Tranfield, David, David Denyer, and Palminder Smart, "Towards a Methodology for Developing Evidence-Informed Management Knowledge by Means of Systematic Review," *British Journal of Management*, Vol. 14, No. 3, 2003, pp. 207–222.

Turella, B., M. Verrocchio, C. Rossi Arnaud, and M. Olivetti Belardinelli, "Children's Memory for Narratives: Influence of Content Familiarity and Input Modality," *Ricerche Di Psicologia*, Vol. 25, No. 4, 2002, pp. 97–115.

Tversky, Amos, and Daniel Kahneman, "The Framing of Decisions and the Psychology of Choice," *Science*, Vol. 211, No. 4481, 1981.

UK Ministry of Defence, *Defence Strategic Communication: an Approach to Formulating and Executing Strategy*, Joint Doctrine Note 2/19, April 2019. As of February 10, 2021: https://assets.publishing.service.gov.uk/government/uploads/system/uploads/attachment_data/file/804319/20190523-dcdc_doctrine_uk_Defence_Stratrategic_Communication_jdn_2_19.pdf

van Laer, Tom, Stephanie Feiereisen, and Luca M. Visconti, "Storytelling in the Digital Era: A Meta-Analysis of Relevant Moderators of the Narrative Transportation Effect," *Journal of Business Research*, Vol. 96, March 2019, pp. 135–146.

Weinstein, Neil, "The Precaution Adoption Process," *Health Psychology*, Vol. 7, No. 4, 1988, pp. 355–386.

Woudstra, Anke Judith, and Jeanine Suurmond, "How Narratives Influence Colorectal Cancer Screening Decision Making and Uptake: A Realist Review," *Health Expectations*, Vol. 22, No. 3, 2019, pp. 330–334.

Zalman, Amy, "Narrative as an Influence Factor in Information Operations," *IO Journal*, Vol. 2, No. 3, August 2010, pp. 4–10.

Zhou, Shuo, and Michael A. Shapiro, "Reducing Resistance to Narrative Persuasion About Binge Drinking: The Role of Self-Activation and Habitual Drinking Behavior," *Health Communication*, Vol. 32, No. 10, 2017.